Geology of the country around Aberdaron, including Bardsey Island

The area described in this memoir lies at the extreme south-western tip of the Lleyn Peninsula (Llŷn). The area is one of the most scenically attractive in Wales, with a dramatic rocky coastline alternating with wide sandy beaches behind which rise cliffs of Quaternary deposits. Inland, several prominent hills of Precambrian and Ordovician rocks rise above lower ground where thick Quaternary glacial deposits occur. The western part of the district is dominated by various rocks of late Precambrian age, whereas the eastern side reveals a cover sequence of Lower Ordovician marine sediments and Ordovician igneous intrusions.

Despite its small size, the area is justly renowned for providing the best exposures of mélange in the British Isles, a spectacular example of a layered igneous intrusion, the remains of the most important manganese mines in Britain and excellent exposures of Quaternary sediments. In addition to these more exceptional geological phenomena, there are other features of interest to both professional and amateur geologists and geographers. These include old jasper quarries and baryte mines, trilobite and graptolite fossil localities, unconformities, complex structures in the Precambrian rocks, examples of rocks sheared and dislocated along fault zones (both cataclasites and mylonites), a calc-alkaline igneous complex, transgressive dolerite sills, peperites and pillow lavas, and Tertiary olivine dolerite dykes.

BRITISH GEOLOGICAL SURVEY

W GIBBONS and
D McCARROLL

Geology of the country around Aberdaron, including Bardsey Island

CONTRIBUTORS
J Bass
J M Horák
A W A Rushton
T P Young

Memoir for 1:50 000 geological sheet 133
(England and Wales)

This memoir, and the 1:50 000 map that it describes, are the product of a contract between the Natural Environment Research Council and the University of Wales College of Cardiff. The interpretations presented are those of the authors.

LONDON: HMSO   1993

© NERC copyright 1993

First published 1993

ISBN 011 884487 3

Bibliographical reference
GIBBONS, W, and MCCARROLL, D. 1993. Geology of the country around Aberdaron, including Bardsey Island. *Memoir of the British Geological Survey*, sheet 133 (England and Wales).

Authors
W Gibbons, PhD
University of Wales College of Cardiff
D McCarroll, PhD
University College of Swansea
Swansea

Contributors
J Bass, BSc
T P Young, PhD
University of Wales College of Cardiff

J M Horák, BSc
National Museum of Wales
Cardiff

A W A Rushton, PhD
British Geological Survey
Keyworth

Other publications of the Survey dealing with this and adjoining districts

BOOKS
British Regional Geology
North Wales (3rd edition) 1961

Mineral Reconnaissance Programme Report
Geophysical and geochemical investigations of the manganese deposits of Rhiw, western Llŷn, North Wales, 1989

MAPS
1:625 000
Great Britain (South Sheet)
Solid geology, 1979
Quaternary geology, 1977
Bouguer anomaly, 1986
Aeromagnetic anomaly, 1965

1:250 000
Cardigan Bay Sheet (25N 06W)
Solid geology, 1982
Sea bed sediments, 1988
Bouguer gravity anomaly, 1980
Aeromagnetic anomaly, 1980

Printed in the UK for HMSO
Dd 295254 C8 2/94

CONTENTS

One Introduction 1
Previous research 1
Regional tectonic setting and outline of the geological history 2

Two Precambrian rocks 6
Gwna Mélange 6
  Clasts within the Gwna Mélange 7
    'Gwyddel Beds' clasts 7
    Basalt clasts 9
    Quartzite clasts 11
    Limestone clasts 11
    Red mudstone clasts 13
    Granitic clasts 14
Sarn Complex 14
  Granite–tonalite 17
  Diorites 18
  Gabbro 19
  Geochemistry 19
Parwyd Gneisses 21
Llŷn Shear Zone 21
  Details 23
Tectonic setting 25

Three Ordovician rocks 27
Aberdaron Bay Group 27
  Palaeontology 29
  Wig Bâch Formation 33
  Porth Meudwy Formation 34
  Aberdaron Bay Group (undivided) 36

Four Intrusive igneous rocks 42
Dolerite 42
Ynys Gwylan 43
The Clip Lava 44
The Mynydd Penarfynydd layered intrusion 44
Olivine dolerite dykes (Tertiary) 46

Five Structural geology 49
Structures in pre-Ordovician rocks 49
  Details 51
Post-Ordovician structures 53
  Folds 53
  Faults 54
    Parwyd fault system 54
    North-westerly striking faults 55
    Daron Fault 55
    Faults in the manganese mining belt 55

Six Quaternary 56
Previous work 56
Preglacial landscape evolution 56
Raised shore platform 57
Evidence of Quaternary glaciation 57
  Erosional evidence 57
Quaternary deposits 61
  Interglacial raised beach 61
  Details 61
  Pre-Late Devensian scree and head 62
  Late Devensian glacial deposits 63
  Details 64
  Interpretation of glacial sediments 69
  Head deposits 70
  Postglacial raised beach 71
  Alluvium 71
  Landslips 71
Quaternary history 72
  Preglacial landscape evolution 72
  Raised rock platform 72
  Quaternary glaciations 73

Seven Economic geology 76
Building stone, jasper, agricultural lime 76
Copper, baryte 76
Sand and gravel 76
Manganese 76

References 78
Appendices
1   1:10 000 maps 84
2   Geological Survey photographs 85
Index 86

FIGURES

1   Geological sketch map of the Aberdaron area 3
2   Sketch map showing the position of the Bardsey district relative to Anglesey and the Welsh Basin 4
3   Outcrop pattern of 'Gwyddel Beds' clasts within the Gwna Mélange of south-west Llŷn 10
4   Microtextures of microgabbros in basaltic clasts within the Gwna Mélange 12
5   Location of granite clasts within the Gwna Mélange in north-west Bardsey Island 15
6   Microtextures in the Bardsey granite clasts 16
7   AFM plot and K_2O v. SiO_2 plot for plutonic rocks from the Sarn Complex 21
8   Box diagram illustrating the structure of the Llŷn Shear Zone at Llangwnnadl 22
9   Sketch structural cross-section across the south-west coast of Llŷn 24
10   Composite lithological log of the Aberdaron Bay Group in the Aberdaron Bay area 27
11   Ordovician graptolites 32
12   Glacial stages in north-west Wales as recognised by Whittow and Ball (1970) 58
13   Evidence for ice-movement directions in western Llŷn 60

14 Quaternary deposits exposed in cliffs east of Aberdaron 66
15 Selected graphic logs of sections exposed in cliffs east of Aberdaron 67
16 Stages of development of the Aberdaron drift sequence 69
17 3-dimensional model of the development of the drift landforms around Aberdaron 71
18 A model which facilitates the incorporation of subglacial sediments at a transition from warm-based to cold-based ice 74

TABLES

1 Representative XRF analyses for the Sarn Complex 20
2 Chrono- and biostratigraphical classification of the Aberdaron Bay Group 31
3 XRF whole rock analyses for lithologies within the Mynydd Penarfynydd layered intrusion 45

PLATES

1 White quartzite clasts in Gwna Mélange on south-west coast of Llŷn 6
2 Fragmented 'Gwyddel Beds' in Gwna Mélange on south-west side of Mynydd Mawr 8
3 Deformed basaltic pillow lavas in a clast within the Gwna Mélange on the south-east coast of Bardsey Island 9
4 Basaltic pillow lavas erupted into micritic sediment within a Gwna Mélange clast on Dinas Fawr 11
5 Bedded pyritic limestone clast within Gwna Mélange on south-west coast between Porth Felen and Pared Llech-ymenyn 13
6 Red mudstone at base of the 'Gwyddel Beds' in cliffs on the west side of Porth Felen 14
7 Mylonitic metasediments within the Llŷn Shear Zone on Mynydd Bychestyn 25
8 View looking west across the coastal exposure of the Aberdaron Bay Group, from the summit of Mynydd Penarfynydd 28
9 Ordovician trilobites 30
10 View west from Bau Ogof-eiral showing south-easterly dipping turbiditic sandstones of the Porth Meudwy Formation 35
11 Overturned slump hook within debris flow in the Porth Meudwy Formation at Bau Ogof-eiral 37
12 Phosphatic ooids within muddy ironstone at Porth Mcudwy 38
13 Volcaniclastic sandstone from immediately above ironstone bed at Porth Meudwy 39
14 Gallt y Mor dolerite intrusion at Porth Ysgo 43
15 Picrite within Mynydd Penarfynydd layered intrusion at Trwyn Talfarach 46
16 Banded melagabbro in Mynydd Penarfynydd layered intrusion 47
17 Thin section microphotographs of olivine dolerite dykes of probable early Tertiary age from the north-west coast of Bardsey Island 48
18 Recumbent F_1 folds in 'Gwyddel Beds' on the coast south of Braich y Pwll 49
19 Upright F_2 folds in 'Gwyddel Beds' on the coast south of Braich y Pwll 50
20 Gwna Mélange east of Porth Felen 53
21 Mynydd Carreg, rising abruptly above plain of thick drift 57
22 Striations and crescentic fractures on quartzite in Gwna Mélange 59
23 Cemented sand and gravel at Porth Oer, originally interpreted as an interglacial raised beach 62
24 Stratified local scree deposits incised by the waterfall of the Ysgo stream 63
25 Glacial drift deposits in cliff section at Aberdaron Bay 65
26 Lower and upper diamicts separated by well-sorted sands at Aberdaron Bay 68
27 Drift overlying weathered pillow lavas at northern end of Porth Oer 70

PREFACE

The area described in this memoir covers the small island of Bardsey (Ynys Enlli) and the adjacent mainland at the south-western tip of the Lleyn Peninsula (Llŷn). This is a remote, unspoilt and beautiful part of north-west Wales which, despite its small size, displays a rich variety of geological features that are of interest to amateur and professional geologists alike. They include Precambrian mélanges, Ordovician layered igneous intrusions and fossiliferous marine sedimentary rocks, Tertiary dolerite dykes, a variety of Quaternary sediments, and old manganese mines. The area also provides a unique link between the Lower Palaeozoic geology of the Welsh Basin and that of Anglesey.

Despite the interest shown in the geology of south-west Llŷn by many illustrious geologists from Adam Sedgwick in the mid-19th century to Robert Shackleton in the 1950s, and the detailed work completed by Charles Matley earlier this century, this memoir, with its accompanying 1:50 000 map, is the first comprehensive account to be published for the district. It describes the structure of the Gwna Mélange and the petrology and geochemistry of the Sarn Complex; and also gives new interpretations of the lithostratigraphy of the Ordovician strata and of the Quaternary geology. Descriptions of the excellent coastal exposures of the solid geology and of the abundant well-defined Quaternary landforms and deposits will make the account valuable to teachers of both geology and physical geography.

Some potential for the exploitation of sand and gravel has been identified in the north-east of the district. The only material of any economic consequence, the manganese ore in the south-eastern corner of the district, was once the most important source of manganese in the UK, although it is no longer worked. However, the mines are still of great interest to the mineralogist and industrial archaeologist.

Peter J Cook, DSc
Director

British Geological Survey
Kingsley Dunham Centre
Keyworth
Nottingham
NG12 5GG

August 1993

ACKNOWLEDGEMENTS

This memoir was mostly written by W Gibbons (solid geology) and D McCarroll (drift geology), following completion of a mapping contract awarded to University of Wales College of Cardiff by the Natural Environment Research Council. The work programme was completed with the co-operation of Dr R A B Bazley and Dr A J Reedman of the British Geological Survey. Contributions to the text were provided by Dr A W A Rushton of BGS (Ordovician faunas), Dr T P Young of UWC Cardiff (Ordovician faunas and lithostratigraphy), Dr J Bass of the University of Oxford (Ordovician sedimentology) and Ms J M Horák of the National Museum of Wales (Sarn Complex). The account was edited by Dr Bazley. The authors also gratefully acknowledge the assistance given by Drs A Beckly, C Harris, F Martin, S Molyneux, R Owens, G Power, G Tegerdine and Professors W T Dean and R M Shackleton.

Special thanks are due to the landowners in the district who always willingly allowed unimpeded access to their land. Diolch yn fawr!

NOTES

Throughout the memoir the word 'district' refers to the area covered by the 1:50 000 geological sheet 133 (Aberdaron, including Bardsey Island).

National Grid references are given in square brackets; all fall within the 100 km square SH.

The authorship of fossil species is given in the index.

The term Llŷn is used throughout for the Lleyn Peninsula; the latter name occurs on the face of the published map.

ONE

Introduction

The area covered by the 1:50 000 Sheet (133) extends from the south-west tip of Llŷn (Penrhyn Llŷn) north-east to near Rhiw and Llangwnnadl. The map also includes Bardsey Island (Ynys Enlli) and the smaller islands of Ynys Gwylan-fawr and Ynys Gwylan-bâch in Aberdaron Bay. It is a remote area which, although attracting a fair number of tourists annually, remains sparsely populated and possesses a landscape that has been little changed by recent human activity. The scenery of the region is characterised by several prominent, rounded hills that rise above low-lying farmland. The scenic appeal of the area is greatly enhanced by a varied, unspoilt and often picturesque coastline that includes several sandy beaches. The wide variety of both inland and coastal scenery is to some extent a reflection of the underlying geology, which is also remarkably varied and interesting. The area includes many types of igneous, sedimentary and metamorphic rocks, and also several sites with especial mineralogical interest. The abandoned Nant and Benallt mines in the south-east of the area have together provided the most important source of manganese ore in Britain.

PREVIOUS RESEARCH

Geological work in the district was pioneered by Adam Sedgwick in the mid-19th century. Sedgwick realised that the rocks found along the north-west coast of Llŷn were similar to the ancient rocks found on Anglesey to the north-west. The first geological maps to be published were produced by the Geological Survey under the guidance of A C Ramsay in the early 1850s; these maps interpreted the oldest rocks on Llŷn to be of Silurian age. This led to considerable discussion and disagreement between geologists in the latter part of the 19th century, with workers such as Hicks (1878, 1879) and Blake (1888) insisting that the supposed 'Silurian' rocks were really Precambrian in age. Blake (1888) coined the term Monian (after the Welsh name for Anglesey: Ynys Môn) to describe these oldest rocks on Llŷn and Anglesey, a name that is still used.

The varied selection of igneous rocks exposed in the cliffline, especially those in the extreme south-east of the district, attracted the attention of eminent 19th-century petrologists such as Bonney (1881, 1885) and Harker (1888, 1889), and a spate of papers on this topic appeared in the 1880s. A particularly useful contribution around this time was made by Raisin, who in 1893 proved that many of the basaltic igneous rocks seen along the west coast displayed pillow texture produced during submarine eruptions. Before the work of Raisin, and others, these basaltic rocks had been referred to as 'serpentines'.

However, the single most important worker on the geology of south-west Llŷn was Charles Alfred Matley, who produced several publications and maps between 1902 and 1939. Although he knew the geology intimately, detailed maps (with the notable exception of Bardsey Island) were unfortunately never published. Among Matley's major contributions were: first, the proof that the Monian rocks were older than 'Silurian' (he believed them to be Precambrian); second, the close correlation of the Monian rocks with similar lithologies mapped by Edward Greenly in Anglesey; and third, the recognition that the younger sedimentary rocks that occur to the south-east of the Monian exposures around Aberdaron and around Rhiw were Lower Ordovician in age (not 'Silurian').

Matley showed that most of the Monian rocks belong to the rock unit called the Gwna Mélange by Greenly (1919). A peculiarity of this unit is that it is completely chaotic, lacking bedding and comprising a mixture of hard rock clasts in a softer slaty matrix. In addition, Matley recognised Monian metamorphic rocks (schists and gneisses) and plutonic igneous rocks (granites and gabbros). He initially believed that these Monian rocks had been faulted against the Ordovician sediments by a major, north-west-dipping thrust (the 'Boundary Thrust'), although he had found an unconformable relationship at two localities on the coast (at Parwyd and Wig). However, as his understanding of the area grew, he retreated somewhat from this initial interpretation. Later work by Robert Shackleton in the 1950s was to reveal yet another locality showing unconformity (at Penrhyn Mawr), which further weakened Matley's 'Boundary Thrust' hypothesis. Shackleton went on to conclude that the Monian/Ordovician boundary was probably unconformable in most places, a view that was maintained by later research workers through into the 1980s. The present survey has shown that whereas the Monian/Ordovician boundary is probably mostly unconformable in the east of the district, the boundary north and west of Aberdaron is a fault.

Following the work of Matley, relatively little detailed research work has been published on the rocks of the area (see reference list). An important paper published by Shackleton in 1956 included the first description of an unconformity exposed beneath the Ordovician at Penrhyn Mawr, a map and an interpretation of the Monian rocks that viewed the plutonic igneous rocks as having been produced from the other Monian lithologies via a prograde metamorphic transition from schists to gneisses. Shackleton also provided a new interpretation of the Gwna Mélange on Llŷn and Anglesey, arguing for an origin as a large-scale sedimentary slide deposit (olistostrome). Schuster (1979, 1980; Wood and Shuster 1978) also concluded that the Gwna Mélange is a regionally developed olistostrome. Although not fully published, Schuster (1980) suggested that the Gwna Mélange is a stratabound sequence with lack of penetrative deformation within the matrix, and provided evidence to show that submarine debris flow and sliding were the dominant transport mechanisms.

The well-exposed igneous intrusion on Mynydd Penarfynydd, originally described by Harker and other workers, excited interest in the late 1960s when Hawkins (1965) recognised it to be a layered intrusion with cumulate textures similar to those being described at the time from areas such as Greenland (the Skaergaard intrusion) and Scotland (the Rhum intrusion). Cattermole (1976) published a geochemical study of these rocks and interpreted them as having been produced by the fractional crystallisation of a hydrated alkali-basalt magma.

The 1980s have seen a resurgence of interest in several aspects of Llŷn geology. The Monian rocks were remapped by Gibbons (1983a, 1989), who recognised a complex structural history within the mélange and, developing an earlier suggestion made by Baker (1969), reinterpreted the schists as defining a shear zone that separated the mélange from the Monian gneisses and plutonic igneous rocks (the latter were taken out of the Mona Complex and renamed the Sarn Complex). Whereas Matley and subsequent workers had believed that the mélange in south-west Llŷn was overlain by a sequence of bedded, cherty sediments called the 'Gwyddel Beds', Gibbons mapped these rocks as merely forming large clasts within the mélange: neither top nor base of the mélange is seen on Llŷn (Gibbons and Ball, 1991). The age of the Monian rocks, although generally believed to be late Precambrian, is not yet definitely known except for the fact that they are clearly older than the overlying Arenig sedimentary rocks. Controversy still exists over defining the absolute age of the Precambrian/Cambrian boundary, a problem exacerbated by continuing uncertainty over the interpretation of radiometric data in north-west Wales (Gibbons and Horák, 1990; Tucker and Pharaoh, 1991). It is therefore not absolutely proven that all the pre-Arenig rocks are Precambrian, because it is possible that some were formed in the early Cambrian, before the deposition of the Cambrian sediments seen on the St Tudwal's Peninsula to the east, in the adjoining Pwllheli district. Despite this uncertainty, present evidence favours a latest Precambrian age for these rocks.

A detailed study of the fossils in the district by Beckly (1985, 1987, 1988) allowed a further refinement of the known ages of the Ordovician rocks, previously recognised as of Arenig and Llanvirn age. Beckly's work showed that in places the Arenig may be subdivided into lower (Moridunian), middle (Whitlandian) and upper (Fennian) stages. This is the only district in Wales where all three subdivisions of Arenig rocks occur in such a small area.

REGIONAL TECTONIC SETTING AND OUTLINE OF THE GEOLOGICAL HISTORY

The rocks of the Aberdaron district and Bardsey Island comprise an ancient basement of mélange, mylonitic schist and calc-alkaline plutons that is overlain by a sequence of Ordovician marine sediments and igneous intrusions (Figure 1). The area lies along the north-west edge of the Lower Palaeozoic Welsh Basin (Figure 2). The basement rocks that form the north-west margin to this basin were probably emergent as a land area during much of the Cambrian Period. They had been structurally deformed during a late Precambrian to early Cambrian phase of tectonic activity along a destructive plate margin. Such activity had produced the blueschists of Anglesey and the Gwna Mélange, and had probably involved a prolonged phase of oceanic plate subduction beneath a continental margin (Wood, 1974). This plate subduction was probably followed by a phase of transcurrent fault movement that moved different geological terranes sinistrally along the continental margin to their present position (Gibbons, 1983b, 1987). Such an event would account for the steep, schistose shear zones seen cutting the Monian rocks and the juxtaposition of units with very different geological histories. Thus the Sarn Complex and Gwna Mélange could represent different terranes moved together from once widely separated areas. The Sarn Complex shows a broad similarity to other inliers of igneous rocks exposed beneath the Cambrian sediments of south-west Wales (e.g. the Johnston Complex) and the Welsh Borderlands (e.g. the Stanner-Hanter Complex). All of these inliers form part of the Avalonian superterrane (Gibbons 1990b), a zone of late Precambrian to earliest Cambrian rocks that continues from central England and south-west Wales through the Avalon peninsula of Newfoundland and down the eastern seaboard of mainland North America. The Llŷn Shear Zone, separating the Sarn Complex from the Gwna Mélange, may therefore represent the north-western margin of the Avalonian area. On the north-west side of this shear zone lie the various units that comprise the Monian rocks of western Llŷn and Anglesey.

To the south-east of the district, thick sequences of Cambrian, Ordovician and Silurian rocks accumulated in the Welsh Basin. Although Cambrian rocks occur in the St Tudwal's Peninsula on the south-east side of Llŷn, there is no evidence that similar Cambrian sediments ever covered the Aberdaron area. Some of the sediments in the St Tudwal's Cambrian succession contain Monian and Sarn Complex fragments and it is likely therefore that the basement rocks were emergent to the north-west at that time.

During Ordovician (Arenig) times the sea encroached north-westwards across North Wales, stepping firstly across the previously uplifted and tilted Cambrian of St Tudwal's and then on over the Sarn Complex and Monian landmass of Llŷn and Anglesey. The water depth of this Ordovician sedimentary basin varied in both space and time, and in North Wales was often strongly influenced by contemporaneous fault movements and volcanic activity. Over 700 m of Arenig marine mudstones, siltstones and sandstones, with minor breccio-conglomerates, were deposited in the Aberdaron area. Tuffs and tuffaceous sandstones provide evidence for nearby volcanic eruptions from the late Arenig (Fennian) onwards. The reason for the volcanic activity is believed to be subduction

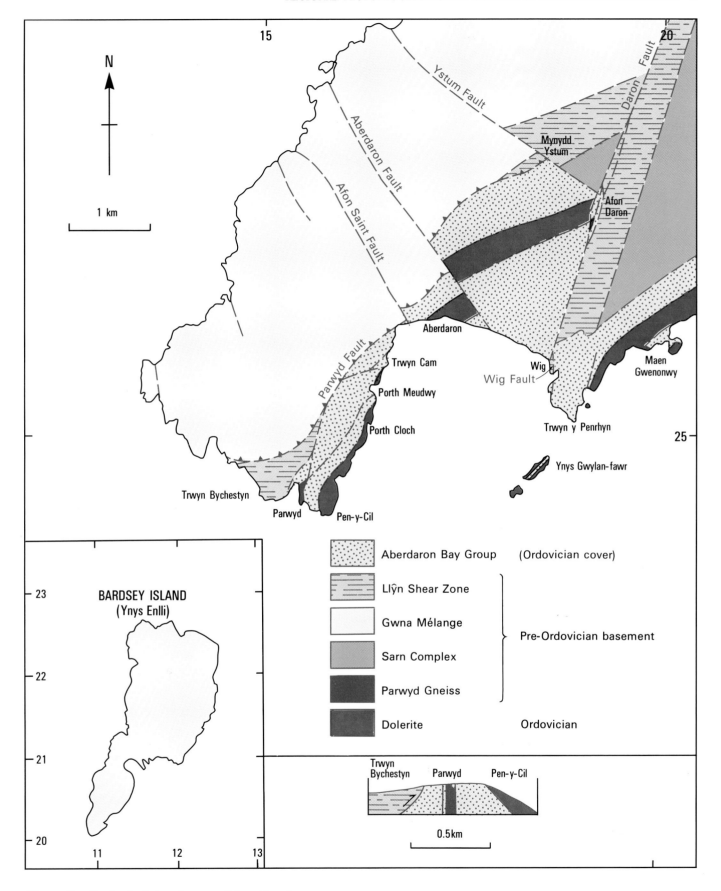

Figure 1 Geological sketch map of the Aberdaron area.

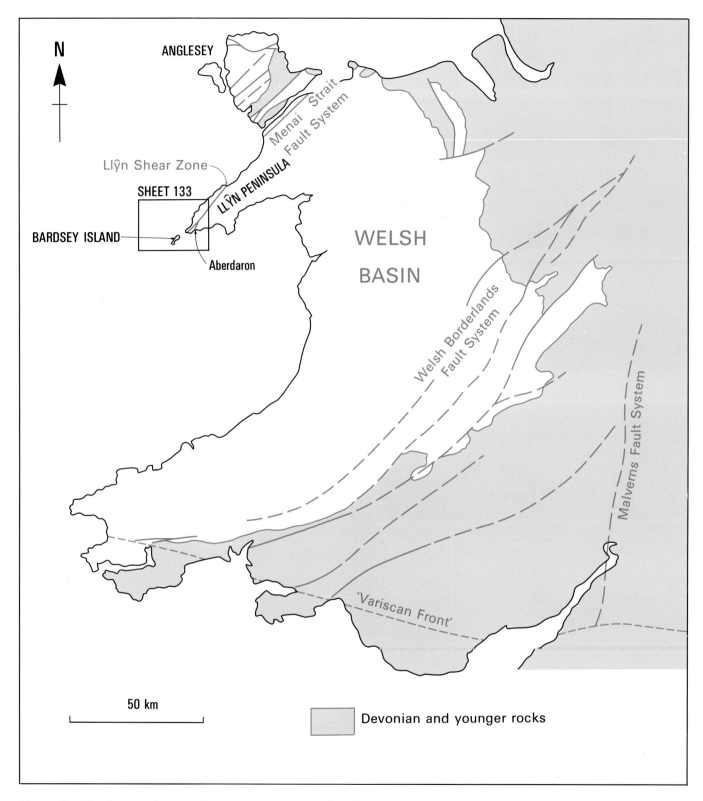

Figure 2 Sketch map showing the position of the Bardsey district relative to Anglesey and the Welsh Basin.

of oceanic crust beneath Wales. Beyond the Monian area, way to the north-west of what is now Anglesey, lay the Iapetus Ocean with a subduction zone active along its south-eastern margin. Final closure of this ocean probably occurred in late Ordovician times (Leat and Thorpe, 1989), followed by the cessation of volcanic activity in the early Silurian. Subsequent uplift of the Lower Palaeozoic Welsh Basin sediments, accompanied by compressional faulting and folding, and the local development of cleavage, occurred in late Silurian to Devonian times.

The Ordovician sediments in the Aberdaron district contain a suite of dolerite intrusions interpreted as being of early Llanvirn age. These are typically sill-like in form and were intruded into wet sediment at shallow levels below the sea floor. Some of these sills display gradations into pillow basalts and basalt–wet sediment mixtures known as peperites. In addition to these basic sills, Llanvirn sediments exposed in the south-east corner of the district contain a thick, layered intrusion. This Mynydd Penarfynydd intrusion represents a thick basaltic sill that differentiated into a succession of rhythmically banded, and probably multiply injected igneous lithologies that range from an ultramafic (hornblende picrite) base to a granophyric top. The southern headland of Mynydd Penarfynydd provides some of the best exposures of layered igneous rocks on mainland UK. This igneous body intrudes the youngest Ordovician sediments (Llanvirn) exposed in the Aberdaron district but shows no evidence for having intruded wet sediment at shallow levels close to the sea floor. The layered intrusion is interpreted as being of probable Upper Ordovician (Caradocian) age, broadly synchronous with the major phase of magmatic activity seen elsewhere in North Wales at around this time.

Although Upper Palaeozoic (Carboniferous and Devonian) sedimentary rocks are preserved in several areas around the periphery of the uplifted Welsh Basin area (e.g. in Anglesey), none occurs on Llŷn. Nor is there any record of Mesozoic rocks, although Cenozoic rocks are preserved as early Tertiary olivine dolerite dykes that intrude the Monian rocks exposed along the coast. These dykes were produced in response to the opening of the North Atlantic Ocean and they represent a distant, minor manifestation of the tremendous magmatic activity centred in Western Scotland and Northern Ireland at this time.

The latest geological events to affect the district, namely the successive advances of glacier ice which overran much of Britain during the Quaternary Period, produced rounded, ice-sculpted hills surrounded by lower areas infilled with thick deposits of glacial tills, sands and gravels. However, the main elements of the present scenery, comprising isolated hills rising above gently sloping plains, are thought to have been produced by a long history of terrestrial weathering and erosion during Mesozoic and Tertiary times. During Pleistocene times the area was crossed by at least three separate glaciers moving south along the Irish Sea basin and producing erosional features such as rock striations and meltwater channels. The thick drift deposits that mantle much of the district are interpreted as having been deposited during the decay of the last (Devensian) ice sheet, when there was widespread downslope redistribution of previously deposited glacigenic materials. Remnants of older Quaternary rocks occur as isolated buried scree and head preserved in sheltered positions beneath the main mass of drift. The most recent drift deposits form small areas of surface head and alluvium.

TWO
Precambrian rocks

The oldest rocks in the district are divided into three main units: the Gwna Mélange, the Sarn Complex and the finely schistose, mylonitic rocks of the Llŷn Shear Zone. The shear zone separates the Gwna Mélange (to the north-west) from the Sarn Complex (to the south-east), although this basically simple relationship has been complicated by later faulting and by an apparent splay of the shear zone to either side of Aberdaron (Figure 1).

GWNA MÉLANGE

This mélange is the most widespread rock unit on the mainland (Figure 1), and forms all of Bardsey Island. It is characterised by a completely chaotic appearance, with jumbled clasts of various rock types in a grey-green slaty mudstone and siltstone matrix (Plate 1). The stratigraphy of these rocks has been disrupted, although on a broad scale it is possible to see changes in the size and types of clasts. Pillow lavas, for example, are particularly abundant as clasts in the mélange exposed around Porth Oer (Whistling Sands) [1670 3015] and east of Porth Felen [1450 2475]. It has been described as a stratabound sequence, characterised by a pseudostratigraphy of various mélange facies, and interstratified with thin coherently bedded units (Schuster, 1980).

The overall thickness of the mélange in the district is probably between 2000 and 3000 m, although this estmate

Plate 1 Large clasts of white quartzite within unbedded, slaty Gwna Mélange in the cliffs forming the south-western tip of Llŷn. Looking east towards Trwyn Maen Melyn [138 253]. (A15005).

lacks precision because bedding is not preserved and the unit has been affected by several phases of folding. Neither top nor base of the mélange are seen, so that this estimate is considered to be a minimum figure. The same mélange is found in eastern, central and northern Anglesey, where it is again several thousand metres in thickness. It is likely that on Llŷn we see only a relatively small remnant of a regional mélange unit once covering thousands of square kilometres.

Structurally, the lowest part of the mélange is exposed along the north-west-facing coast in the north-east corner of the district. Here, most of the clasts within the green slaty matrix are various types of sandstone, locally mixed with less common lithologies such as limestone, white quartzite, basaltic lava and red mudstone. Between just south of Penrhyn Melyn [2070 3550] and Porth Ysgaden [2215 3757], most clasts are derived from a distinctive feldspathic sandstone sequence. This illustrates how, despite the chaotic texture of the mélange, it is possible to subdivide the rocks into various facies depending upon the relative abundance of clast types.

Moving south-west along the coast from the inlet south of Maen Aber-dywyll [1893 3342], the mélange becomes much richer in dark purple and green, commonly pillowed basaltic lava. Many of these lava clasts are hundreds of metres in size, such as that seen in Porth Widlin [1830 3246], and they form a prominent belt running for 3 km up the coast from Porthorion [1557 2860] to north of Trwyn Glas [1658 3105], and thence inland for several more kilometres towards the north-east. Locally, as in Porth Widlin and inland at Pen-yr-Orsedd [1915 3168], the basalts contain intrusive masses of microgabbros. The most prominent body of basic lava forms the hill of Mynydd Carreg [1630 2915], where jasper has been quarried. Clasts of jasper, limestone and other lithologies are mixed up with the basic lavas. In some places the lava is intimately intermixed with limestone, and can be seen to have erupted into a limestone mud before becoming disrupted within the mélange, for example on the coast at Dinas Bâch [1575 2935] and Dinas Fawr [1550 2905] (Plate 4).

The distinctive and colourful pillow lava mélange gives way south-westward to an overlying and very different mélange facies. The latter is dominated by clasts of commonly coarse, feldspathic, locally tuffaceous sandstones and, distinctively, by masses of bedded, cherty siltstones and sandstones. These lithologies are mixed with white quartzite, limestones, red mudstones, dark basaltic lavas and other rocks to produce a colourful chaotic mixture. The cliffline exposures of these rocks on the south-west coast around Trwyn Maen Melyn are renowned as providing the best, most easily accessible examples of mélange in the UK (Plate 1). The cherty siltstone and sandstone clasts were called the 'Gwyddel Beds' by Matley (1928) (type locality: Mynydd y Gwyddel [1420 2520]) who believed them to lie above the Gwna Mélange. In fact, these distinctive, white-weathering, fine-grained sediments are part of the Gwna Mélange (Plate 2), although in places they form huge clasts that have retained original bedding (Gibbons and Ball, 1991). The largest of these clasts forms most of the hill on the west coast known as Mynydd Anelog, which rises to a height of 191 m. Other large masses of these cherty sediments occur on Mynydd y Gwyddel [1420 2520] and on the slopes of Mynydd Mawr [1360 2550].

The commonest lithology in the clasts of the 'Gwyddel Beds' is a finely laminated, cream-weathering, cherty rock with a subconchoidal fracture; fine-grained, locally cross-bedded sandstones also occur. Red siliceous mudstones are interbedded with these cherty rocks at the base of several clasts; in these, the original stratigraphical sequence is preserved. However, in many places this stratigraphical relationship has been destroyed by disruption during mélange formation. The original relationship is well preserved on the west side of Porth Felen [1437 2497], around Uwchmynydd [e.g. at 1550 2620] and south of Braich y Pwll [1353 2587]. The local incorporation of the red sediments into the mélange is shown by exposures at Braich y Pwll and, more especially, along the steep coast to the north-east [e.g. at 1385 2608]. Elsewhere in the south-west, clasts of cherty 'Gwyddel Beds' material in the mélange are common in many areas, and they occur sporadically along the coast as far north-east as Porth Bâch [1735 3214].

The whole of Bardsey is composed of Gwna Mélange, with the island being divisible into two broad sub-areas. The outcrop on the eastern side of the island, including the one prominent hill of Mynydd Enlli, is very similar to the exposures on the mainland immediately to the north-east. This mélange type is characterised by a chaotic mixture of white quartzites, limestones, cherty sediments ('Gwyddel Beds'), basic lava and many different, commonly coarse-grained and feldspathic sandstones. These clasts all lie in an unbedded slaty matrix. By contrast, the mélange exposed on the western side of the island is generally dominated by disrupted, finer-grained greywacke sandstones, siltstones and mudstones. This latter mélange also contains areas rich in limestone and quartzite, notably at Pwll Hwyaid [1118 2083], Trwyn Dihirid [1128 2118], immediately west of the lighthouse [1105 2063] and east of Cefn Enlli [1180 2103]. The major contrast between these two mélange types, however, is in the nature of the sandstone clasts. The eastern mélange, dominated by coarse, feldspathic sandstones, overlies and dies out south-westwards into a mélange characterised by disrupted slabs of generally finer-grained, grey greywacke sandstone. Two other features are noteworthy within the Bardsey mélange. Firstly, the presence of a large slab of deformed basic pillow lava (Plate 3) and lava-limestone breccia south of Hellwyn [1140 2047]; and secondly, the existence of granitic clasts along the north-western and western coast.

Clasts within the Gwna Mélange

'Gwyddel Beds' clasts

The term 'Gwyddel Beds' was introduced by Matley in 1928 to describe a series of fine-grained, cherty, bedded sediments exposed in the extreme south-west of Llŷn with Mynydd y Gwyddel [142 252] being the type locality. Large masses of these rocks form most of each of the prominent hills on the mainland of south-western Llŷn,

Plate 2 Fragmented 'Gwyddel Beds' in Gwna Mélange on south-west side of Mynydd Mawr [1385 2550].

namely Mynydd y Gwyddel, Mynydd Mawr (Figure 3) and Mynydd Anelog. Similar rocks occur on Mynydd Enlli on Bardsey Island [1255 2215], and are also found as large masses within the Gwna Mélange at Ogof Newry [1650 3113], Porth Llong [1665 3088], Porth y Wrâch [1674 3073] and south of Porthorion [1547 2859]. The commonest lithology weathers to a distinctive, finely laminated, cream coloured, cherty rock with a subconchoidal fracture. Microscopically, the rock is composed almost entirely of a microcrystalline siliceous groundmass within which lie small crystals of quartz, feldspar and white mica. Most other beds are fine-grained feldspathic litharenites containing subrounded to angular grains of quartz, K-feldspar, perthite, plagioclase (oligoclase-albite), muscovite and rare epidote, and rare fragments of decomposed basalt, all mixed with intraclasts of the cherty lithology and set in a fine siliceous matrix. These sandstones in places exhibit small-scale cross-bedding (typically in the upper part of the bed) and are best exposed on the coast [1352 2553], south of Braich y Pwll. There is a complete gradation between these two end-member lithologies: the sediments are bimodal, with a fine siliceous matrix forming a background to varying inputs of clastic detritus. There are also thin brown or buff shaly mudstone beds composed of cryptocrystalline silica and tiny flakes of white mica.

Matley (1928) considered the cherty lithology to be of pyroclastic origin (air-fall volcanic dust), and Shackleton (1956) suggested that it represented acid tuffs. These interpretations are supported by the angularity of some quartz grains, although obvious pyroclastic textures are lacking. The volcanic component of these rocks has been reworked by sedimentary processes and mixed with epiclastic detritus derived from a mixed provenance that included both granitic and basaltic material.

The mélange that surrounds the larger 'Gwyddel Bed' clasts contains a mixture of sandstones, smaller fragments of 'Gwyddel Beds', limestones, basalts, quartzites, jasper and red mudstone. All of these lithologies lie within the usual grey-green, semipelitic matrix. The best exposures of this mélange facies occur along the precipitous coast south-west of Ogof Pren-côch [1541 2838], where it contains a prominent train of large, lenticular slabs of feldspathic sandstone [e.g. 1530 2837; 1515 2822], a large mass of 'Gwyddel Beds' [1519 2827] and several clasts of basic lava [1515 2825 to 1515 2815]. Away from the coast, on the north-facing slopes of Mynydd Anelog, the mélange clasts are mostly pale-weathering,

Plate 3 Deformed basaltic pillow lavas in a clast within the Gwna Mélange on the south-east coast of Bardsey Island (Ynys Enlli) [1136 2045].

fine-grained, cherty 'Gwyddel Beds', with a particularly prominent mass showing sinistral offset by faulting [1519 2819]. By far the largest mass of 'Gwyddel Beds' forms the summit and southern slopes of Mynydd Anelog. The summit area is poorly exposed, but good exposures may be found at several places on the hillside [e.g. 1555 2647], where the rocks show their original bedded character.

Basalt clasts

Numerous masses of dull green to purple basaltic volcanic rocks are common within the Gwna Mélange. Many of these display pillowed textures (the 'spheroidal structure' of Raisin, 1893) and crop out as lenticular slabs that have remained upward facing during dismemberment within the mélange (Plate 6). The best exposures occur along the 1.5 km stretch of coast from the south side of Porth Oer [1635 2978] to just south of Porthorion [1555 2856]. Disrupted masses of dark green basic lava, commonly pillowed, are mixed with pale grey limestones, red mudstones, jasper, sandstones, and cherty 'Gwyddel Beds', all of which lie within a slaty, grey-green, semipelitic mélange matrix. Large clasts of pillowed lava flows are abundant and often extremely well preserved, for example on Dinas Bâch [1577 2938], where individual SSE-facing flows are coated with rubbly breccia. A striking feature of many exposures along this coastline is an intimate mixing of lava and limestone, produced by eruption of the lavas into wet micrite (Plate 4). Lava-limestone eruptive breccias are well exposed on the south side of Dinas Fawr [1555 2907] where they are associated with NNW-facing pillow lavas. Where limestone is locally abundant within the lava mélange, complex chaotic and veined textures have developed, with carbonate partially replacing the altered lava. The limestones occur between individual pillows, as fragments in colourful lava-limestone breccias, and as almost pure limestones within which lie embedded lava fragments.

Inland exposures of jaspery lavas, apart from the jasper quarry on Mynydd Carreg, are few, but include occurrences on either side of the stream around Pont Cyll-y-felin [1742 2833; 1749 2857]. Clasts similar to the basaltic lavas exposed on the coast around Porth Oer are found inland north-east for 3 km to Pen-yr-Orsedd [1915 3170]. Similar exposures occur a further 2km north-east from Pen-yr-Orsedd to Llangwnnadl [2092 3320]. At two localities, one on the coast at Porth Widlin [1830 3247], the other inland at Pen-yr-Orsedd [1915 3170], a distinctive pale green microgabbro intrudes the basaltic lavas. The pri-

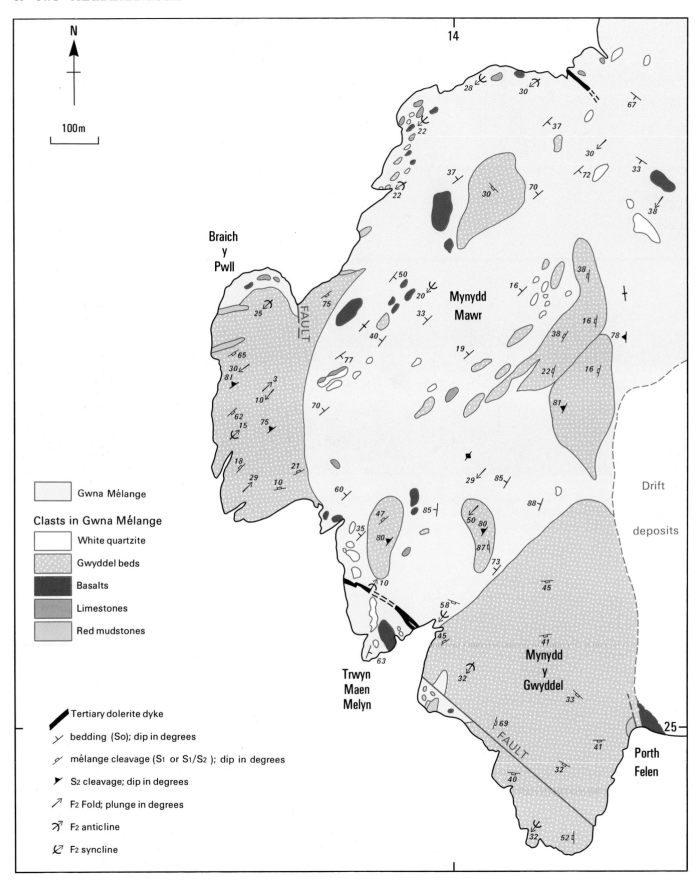

Figure 3 Outcrop pattern of 'Gwyddel Beds' clasts within the Gwna Mélange of south-west Llŷn.

Plate 4 Basaltic pillow lavas erupted into micritic sediment within a Gwna Mélange clast on Dinas Fawr [1565 2904].

mary mineralogy of this rock is extensively replaced by secondary chlorite, epidote, calcite, white mica and actinolite, and is commonly strongly sheared (Figure 4).

Quartzite clasts

Although forming only a small part of the Gwna Mélange, clasts of white orthoquartzite provide one of the most conspicuous characteristics of it (Plate 1). The clasts are usually 0.5 to 5 m in length and commonly occur in clusters and trails associated with limestone lenses. Exceptionally prominent masses, referred to by Matley (1928) as 'The Great Quartzite', are exposed on the north-west side of Trwyn Bychestyn [1490 2445], at Porth Tŷ-mawr [1885 3315] and at Careg y Defaid [1938 3435]. Prominent clusters of white quartzite lenses crop out around Porth Colmon [1950 3433], east of Porth Ysgaden [2246 3746] and in the mélange on Mynydd Mawr [e.g. 1375 1570].

Limestone clasts

Limestone clasts are common within the mélange. The main types are white to grey limestone, pinkish, brown-weathering dolomitic limestones, and the lava-limestone breccias associated with basaltic clasts. White limestones occur as long, thin masses within the grey-green slaty mélange matrix, in places intimately interleaved with the matrix to produce a streaky, marble-like appearance ('slaty marble mélange'), for example on the coast west of Porth Colmon [1930 3434 to 1920 3421]. Blue-grey, calcite-veined limestone clasts occur as pods, commonly associated with brown dolomitic limestone and white quartzites. In places these pods are large, over 10 m in length, and weather out to produce deep solution pits in the cliffs. A good example of grey limestones is to be found on the north-west coast of Bardsey Island [1148 2247]. The largest exposed limestone clast within the mélange in this district is over 65 m long and occurs as a bedded, brown, pyritic limestone running along the south-west coast near Pared Llech-ymenyn [1460 2470] (Plate 5). This limestone occurs immediately above a massive block of jaspery pillow lavas and below a prominent thrust plane. The limestones associated with basaltic clasts are usually either brown-weathering dolomite or pink, fine-grained, manganiferous, partially dolomitised limestone. Other recommended localities for examining the limestones associated with lavas are on the coast around Dinas Fawr [1556 2905], Trwyn Maen Melyn [1385 2515] and north-west of Pared Llech-ymenyn [1460 2469].

Figure 4 Microtextures of microgabbros in basaltic clasts within the Gwna Mélange. (For details, see opposite.)

Plate 5 Bedded, pyritic limestone clast within Gwna Mélange on the south-west coast between Porth Felen and Pared Llech-ymenyn [1460 2470].

Figure 4 Microtextures of microgabbros in basaltic clasts within the Gwna Mélange.

A. Primary igneous texture is preserved as euhedral and subhedral plagioclase crystals (stippled) enclosed by poikilitic clinopyroxenes. The latter have been entirely altered to fine-grained green aggregates of chlorite, epidote, actinolite, sphene and iron oxide. Primary crystals of ilmenite are altered to leucoxene (dark brown). Plagioclase crystals are sericitised and extensively fractured. The small vein (centre) is filled with small, fresh albite crystals. Pen-yr-Orsedd [1915 3170].

B. Cracked and chloritised augite phenocrysts display remnants of subophitic texture (right). The rock is cut by a mylonitic foliation within which broken augite phenocrysts (left) lie as partly chloritised porphyroclastic augen embedded in a fine-grained matrix of white mica and epidote derived from the alteration of plagioclase. Small fibres of actinolite lie within the chlorite. Beach west of Penrhyn Nefyn [2920 4095].

C. Mylonitic texture: porphyroclasts of altered plagioclase and pyroxene set in a fine-grained foliated matrix composed of epidote, iron ore, white mica, chlorite and sphene. Porth Widlin [1830 3247].

Scale bar = 0.1 mm

Red mudstone clasts

Exposures of red, shaly, hematitic mudstones, although relatively rare, are conspicuous. They are associated with basaltic lava clasts and 'Gwyddel Beds' clasts, as red phyllite metasediments within the Llŷn Shear Zone and as individual fragments in sandstone clasts within the mélange. Red shaly mudstones occur within the colourful and chaotic lava-limestone mixtures in Porth Felen [1444 2495] and north of Porthorion [1560 2888]. Within the 'Gwyddel Beds' clasts there are two red mudstone beds, one at the base of the sequence and consisting of thin interbeds of red mudstone within pale cherts and rare black mudstones. There is a gradation from this red mudstone into overlying siliceous 'Gwyddel Beds' lithology, the best exposure of which can be seen on the west side of Porth Felen [1370 2497] where approximately 6m are visible (base not exposed) (Plate 6). The other red mudstone unit is found in bedded contact with and within the 'Gwyddel Beds' only on the coast south of Braich y Pwll [1353 2582]; it lies about 15 m above the basal red mudstone. This second, younger mudstone sequence is up to 8 m thick and contains fewer seams of pale cherty material than the lower one. The same red mudstones occur inland around the southern side of

Plate 6 Red mudstone exposed at base of the 'Gwyddel Beds' in cliffs on the west side of Porth Felen [1437 2497]. The rocks are folded by south-westerly plunging upright folds with an associated subvertical axial planar cleavage (S_2).

Mynydd Anelog and are especially well exposed in the farmyard at Ystol-helig-bâch [1552 2619], beneath the cottages of Bryn-Chwilog [1540 2628] and along the track running north-west from Capel Uwch-y-Mynydd [1542 2638].

Most exposures of red mudstone occur as dissociated rafts and smaller fragments disseminated within a green slaty mélange matrix. Of the numerous exposures, some of the best occur in the cliffs around the south-western tip of Llŷn, for example at Porth Felen [1445 2695], Mynydd y Gwyddel [1395 2505] and Mynydd Mawr [1389 2648]. At the last locality there are slabs of the banded mudstone interbedded with cherts, which are seen undisrupted on the west side of Porth Felen. In some places, notably on the east side of Porth Felen [1445 2695], on the west side of Pared Llech-ymenyn [147 246] and the east side of Bardsey Island [1206 2092], the red mudstone has become the semipelitic mélange matrix within which lie more competent clasts of sandstone and other lithologies. Fragments of 'Gwyddel Beds' are commonly associated with such red mudstone mélanges.

Granitic clasts

Clasts of granitic rocks are exceptionally rare in the Gwna Mélange in north-west Wales, the only known localities being on the west side of Wylfa Head in northern Anglesey (Greenly, 1919, p.307) and on Bardsey Island, where lenticular masses of granitic rocks are exposed on the north-west coast [1153 2212 to 1172 2257] and in Porth Solfach [1140 2125]. The granitic rocks were discovered by Raisin (1893) and described by Matley (1913) who suggested that the exposures may represent an intrusive sill repeated by folding. However, the main granitic masses are all rootless, forming sheared lensoid clasts in equally sheared disrupted sediment. These clasts are up to about 120 m in length and about 50 m in width, and do not show intrusive chilled margins (Figure 5). Those exposed in the north-west corner of the island are all pale muscovite granites, although those in Porth Solfach and the large mass on the north-west coast west of Bae y Rhigol [1177 2257] are of more heterogeneous composition, and include both granitic and more mafic lithologies.

The granite clasts are characterised by pervasive cataclastic textures (Figure 6). Exposures display a foliation parallel to and gradational into the S_1 fabric present in the surrounding mélange. Imposition of this fabric produced mylonitic textures within the granitic rocks, and its development was accompanied by pervasive low-grade alteration of the primary mineral phases. The rocks show all gradations from unfoliated cataclasites to thoroughly foliated, but low temperature mylonites. The deformation mechanisms that produced the dynamic shearing fabric included both ductile processes (especially the dynamic recrystallisation of quartz) and brittle cataclastic fracture and flow. These indicate that the process responsible for the generation of the Gwna Mélange involved considerable cataclastic deformation of crystalline clast lithologies such as these granites. A general discussion on the tectonic setting of the Gwna Mélange is given below (p.25).

SARN COMPLEX

The mostly granitic to dioritic (locally gneissic) rocks of the Sarn Complex occur to the east of the Gwna Mélange outcrops and are very poorly exposed. Inland exposures of the Sarn Complex comprise a range of plutonic igneous lithologies from granite to gabbro, the granitic rocks being more common than the mafic lithologies. The most north-easterly exposures of the Sarn Complex, on the flanks of Mynydd Cefnamwlch [227 339], are of a homogeneous, pale, adamellitic granite known as the Sarn Granite. South-west of this hill, dioritic rocks are exposed at Carrog [2175 3305] and along the Llangwnnadl stream [2120 3277]. These diorites exhibit a range of textures from unfoliated, primary igneous (at Carrog) to foliated and gneissic (as along the Llangwnnadl stream). The dioritic rocks are cut by unfoliated granite veins and

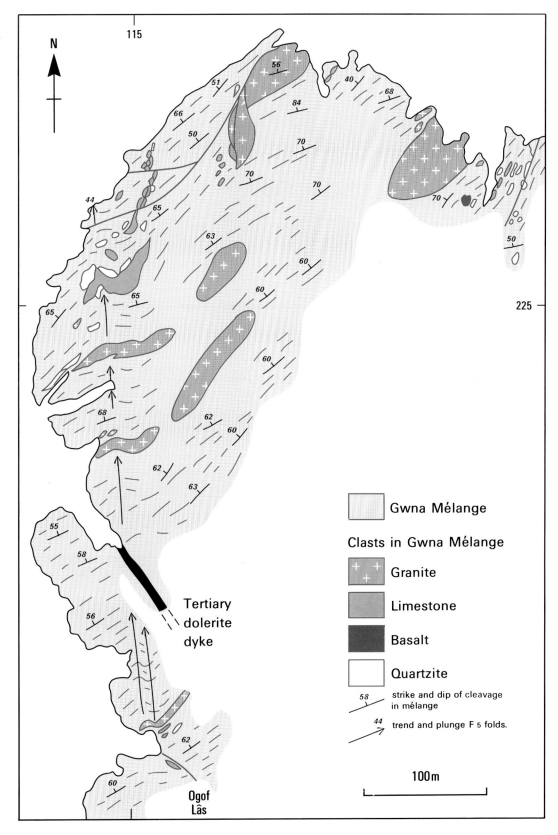

Figure 5 Location of granite clasts within the Gwna Mélange in north-west Bardsey Island.

16 TWO PRECAMBRIAN ROCKS

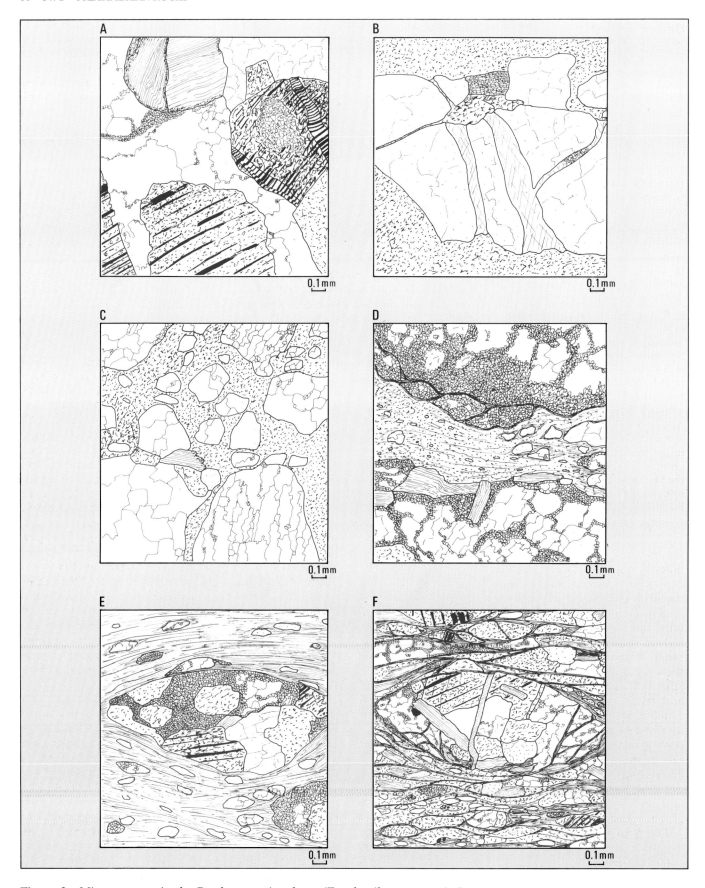

Figure 6 Microtextures in the Bardsey granite clasts. (For details see opposite).

Figure 6 Microtextures in the Bardsey granite clasts.

A. Modified granite texture. Quartz (white) shows extensive subgrain development and primary recrystallisation along serrated grain boundaries. Muscovite (top left) shows kinked cleavage and marginal recrystallisation to white mica. Plagioclase (heavy stipple) is deeply sericitised: the grain on the right displays two crack seals, each infilled with clear plagioclase, and a central patch of primary recrystallisation K-feldspar (top right) displays subgrains, but is less strained than quartz. North-west tip of Island [1163 2265].

B. Broken, strained orthoclase crystal set in sericitic matrix (stippled) derived from the decomposition of plagioclase. Orthoclase fragments separated by calcite vein infill. Quartz grain (top right) is completely recrystallised.

C. Cataclastic microbreccia of strained quartz, K-feldspar and muscovite fragments set in matrix of fine quartz and white mica.

D. Weakly foliated microbreccia zone (centre) between extensively recrystallised parent (muscovite granite). Tiny new quartz grains grow from subgrain boundaries.

E. Recrystallised and sericitised granite fragments in foliated pelitic mélange matrix surrounding the granite clast from which B, C and D were drawn. The granite fragments have become dissociated from the larger granite clast and mixed with clastic quartz and feldspar grains set in an unbedded pelitic matrix. B, C, D and E all from the north side of Ogof Las [1153 2212].

F. Mylonitic foliation in granite. Augen of parent granite, showing the deformation textures seen in A to E, separated by an inosculating network of shear planes. In much of the rock the granite texture has been replaced by fine-grained intergrowths of white mica, chlorite and opaque ore derived from the breakdown of plagioclase and biotite, and small crystals of dynamically recrystallised quartz and orthoclase. West side of Bae y Rhigol [1177 2257].

Scale bar = 0.1 mm

dolerite dykes. At the southern end of the Sarn Granite exposures, around the low hill of Pen y Gopa [223 317], there are granitic, granodioritic and dioritic rocks. These mixed, often xenolithic, hybridised rocks are rich in biotite, but locally contain hornblende and clinopyroxene. Hybrid textures of this nature are characteristic of most other inland exposures of the Sarn Complex, notable those near Crugan Bâch [2125 2995] and Meillionydd [218 292]. The latter locality provides one of the largest and freshest exposures of Sarn Complex lithologies, which range from diorites through to granites, and which locally show a foliation. The only coastal exposures previously grouped with the Sarn Complex are along the east side of the precipitous cove known as Parwyd [154 243]. These exposures, however, are of basic, garnetiferous gneisses (Parwyd Gneisses) that are quite different from and possibly unrelated to typical Sarn Complex lithologies; they are considered below (p.21).

The Sarn Complex represents a phase of pre-Arenig magmatic activity and shows a petrology and geochemistry compatible with the rocks having crystallised within a subduction-related arc environment. A Rb-Sr isotopic date of 549±19 million years (Beckinsale et al., 1984) probably provides a minimum age for these rocks. The Sarn Complex is broadly similar to other mixed intermediate to acid plutonic rocks that occur within the late Precambrian basement of southern Britain, such as the Johnstone diorites of Pembrokeshire. However, it remains uncertain whether correlation between the Sarn Complex and the basement exposed on the opposite side of the Welsh Basin is justifiable. An alternative correlation is with Monian granitic rocks such as the Coedana Granite in Anglesey. Another possible relationship, as yet unproven, is between the Bardsey granites and the Sarn Complex.

Granite–tonalite

The largest, most homogeneous granitic mass in the Sarn Complex is the Sarn Granite, the type locality of which is the prominent rounded hill of Mynydd Cefnamwlch [227 339], only the south-western end of which is present in the district. The Sarn Granite is generally poorly exposed, but there is a good exposure in an abandoned quarry beside the B4417 on the west side of Mynydd Cefnamwlch [222 339].

Early work on this granite was chiefly concerned with its age. The rock was initially mapped as an 'intrusive feldspathic porphyry' of probable Cambrian age (Ramsey, 1866). Hicks (1879) called it the 'Rhos Hirwain Syenite' and placed it within the Precambrian. Tawney (1883) examined the contact between the 'so-called syenite' and Palaeozoic sediments at Mountain Cottage Quarry [230 347] and tentatively suggested a faulted relationship. A more detailed description was provided by Harker (1888) who identified the rock as a granite rather than a syenite. An intrusive contact with Arenig sedimentary rocks was suggested by Matley (1932), but in 1935 he revisited Mountain Cottage Quarry, this time with Smith, who recognised an Arenig basal conglomerate containing granite pebbles (Matley and Smith, 1936). The granite had therefore been proven to be pre-Arenig, and was compared to the Monian granites on Bardsey Island and Anglesey (Coedana granite). Shackleton (1956) interpreted the granite as the final anatectic product of a prograde transition from low-grade Gwna Mélange, through schists and gneisses, to granite. Later work by Baker (1969) and Gibbons (1983a) reinterpreted this supposed 'prograde transition' as a ductile shear zone that separates the granitic and gneissic rocks from low-grade Gwna Mélange.

To the south of the Sarn Granite, other exposures of the Sarn Complex reveal more varied lithologies, and it is likely that there is a gradational boundary between the Sarn Granite and other components of the Sarn Complex. Important, though now badly degraded exposures of the latter occur around the low hilltop at Pen y Gopa [223 317]. Most of them are of granite and granodiorite, and Matley (1928) mapped them as part of the Sarn Granite. However, Shackleton (1956), recognising the heterogeneity of several exposures around the hill, described them as 'streaky and xenolithic tonalites and adamellites...with streaky or faint foliation...mineralogically very similar to the Sarn Granite'. The most mafic rock types occur in the track on the hilltop [223 318]. They contain dark amphibolitic inclusions and are irregularly veined by granite and pegmatite. These exposures provide a connection between the granitic rocks of the Sarn Granite and the more basic exposures to the west and south. Near Four Crosses [213 310] an unsuccessful well sinking made in 1921 passed through 6 m of glacial till into 2.4 m of granitic rock (Matley, 1928). A thin section shows a heterogeneous granitic rock similar to that exposed at Pen y Gopa.

Small exposures of heterogeneous granitic rocks also occur near Crugan Bâch [2125 2995]. They were briefly described as 'acid gneiss' by Matley (1928) and as 'acid rocks with streaky or faint foliation' by Shackleton (1956). Gibbons (1980) noted the presence of foliated amphibolitic xenoliths, locally overprinted by cataclastic textures. Similar heterogeneous granitic to dioritic rocks are exposed in an old quarry at Cadlan-uchaf [198 267].

Some of the best exposures of the Sarn Complex occur around the hilltop east of Meillionydd Farm [217 292]. They are mostly granitic and are variably foliated, as noted by both Blake (1888) and Harker (1888), who described the rocks as 'granite and gneissose granite'. Gibbons (1980) described the rocks as including 'hornblende-biotite granodiorite' and 'gneisses' cut by granite. The northern part of this outcrop is tonalitic and has abundant inclusions of foliated granite gneiss, whilst the southern end is more granitic in composition and contains conspicuous inclusions of fine-grained basic material. These inclusions are considered to be the same as those observed within the diorite at this locality but, in contrast to the latter, they are generally rounded in form and several centimetres in diameter, and have sharp contacts with the host rock. A weak foliation is found in many of these inclusions, which suggests that they represent the same material as is seen at Crugau Bâch, Pen-y-Gopa and Cadlan.

Acid to intermediate rocks of the Sarn Complex have equigranular textures, with quartz and feldspar showing igneous textures modified by subsolidus recrystallisation, particularly in the most acid members. Classifications using both the modal mineralolgy and the normative mineralogy (Streickeisen, 1976) have been employed, although such classifications are based on proportions or compositions of feldspars, which are prone to alteration, and therefore give only an approximation to the primary mineralogy.

Mineralogically, the transition from tonalite to granite is characterised by a decrease in ferromagnesian phases, with a reciprocal increase in feldspar and quartz, and an increase in potassium feldspar relative to plagioclase. Ferromagnesian phases within the tonalite at Mellionydd are dominated by clinopyroxene, rimmed by hornblende, although biotite is also present. In the granodiorite, biotite is dominant over green hornblende; it may constitute up to 15 modal per cent and occurs both as an interstitial phase and rimming hornblende. In granitic rocks biotite is the only ferromagnesian phase present, but is commonly altered to chlorite plus either discrete sphene grains or intergrowths of sphene and oxides; where unaltered, it is foxy red. Pleochroic halos around zircons are preserved within amphibole or, more commonly, chloritised biotite; individual zircon crystals up to 0.05 mm in length are conspicuous in the granodiorite. Apatite is present in all compositions, but is most abundant in the granodiorite and relatively scarce in the granite. No other accessory phases have been identified. Oxide phases are restricted in occurrence to secondary magnetite. Subhedral to anhedral plagioclase is the most abundant phase in intermediate and acid rocks; it shows polysynthetic albite twinning, although simple Carlsbad twins are also common. Compositions range from andesine in tonalite to oligoclase-albite in granite, but zoning has not been identified optically. Recrystallisation of feldspars has resulted in crenulate margins to many crystals. Microcline is present as both twinned and optically homogeneous forms, whilst perthites contain impersistent albite stringers. Both potassium feldspar and plagioclase are altered to fine mats of white mica. Graphic intergrowths of microcline and quartz are present within late-stage granophyric veins from Pen y Gopa; this texture is also locally developed within the main body of the Sarn Granite.

Diorites

Diorites occur as several isolated exposures lying south of the Sarn Granite outcrop, the best of which are at Llangwnnadl and Carrog. They vary from well foliated rocks to massive, coarse-grained rocks with unfoliated plutonic texture. Foliated varieties are best seen along the banks of the Afon Fawr for 600 m downstream from Pont Llangwnnadl. The exposures at Llangwnnadl have been variously described as 'gneissic diorite' and 'massive diorite' (Harker, 1888), 'hornblendic gneisses' (Matley, 1928), 'hornblendic migmatites' and 'amphibolites' (Shackleton, 1956) and 'foliated diorite' (Gibbons, 1980). They are amphibolites with a gabbroic or dioritic protolith. The nearby exposures at Carrog [2175 3300] lie some 600 m north-east of Pont Llangwnnadl, but have received relatively little attention from previous workers. Matley initially mapped them as 'gabbro' but then changed this to 'gneisses'. The Carrog exposures are more obviously igneous in texture, being essentially unfoliated and of plutonic appearance. It is interesting to note that the only Sarn Complex exposures to show a well-foliated texture occur at Llangwnnadl, adjacent to the Llŷn Shear Zone (p.22). It is possible therefore that the Llangwnnadl amphibolitic rocks were produced in response to early, high-temperature shearing along this terrane boundary.

The dioritic rocks locally show a compositional range towards the gabbroic lithologies at Graig Fael (see below). This transition from gabbro to diorite is effected by a decrease in the anorthite component in plagioclase, the occurrence of olive-green magnesio hornblende in place of brown titanium magnesio-hornblende and the appearance of ilmenite. Clinopyroxene ($Ca_{46}Fe_{15}Mg_{39}$) in the diorites is of similar composition to that in the gabbro, but is found only as irregular inclusions within hornblende, and is less abundant. Textures from the Carrog diorite show a framework of euhedral plagioclase, typical of a cumulate phase. Clinopyroxene is also interpreted as a cumulate phase, occurring as ovoid bimineralic crystal accumulates with ilmenite within the diorite. Magnesio-hornblende, quartz and perthite constitute the intercumulus phases, which occur both as simple infills of intercumulus spaces and also, in the case of hornblende, as rims to cumulus clinopyroxene. Apatite is present as inclusions within all phases except clinopyroxene, whilst ilmenite is most commonly associated with or included within the ferromagnesian phases. The high modal abundance of both apatite and ilmenite suggests that these are likely to have been concentrated by accumulation. Diorites from Llangwnnadl have the same essential mineralogy as described above, although ilmenite may be partially or totally replaced by sphene, and clinopyroxene may be absent. Other textures, such as fine mosaic intergrowths of quartz and hornblende, suggest that these diorites may have experienced subsolidus recrystallisation, which has obliterated or modified primary igneous textures.

Other exposures of dioritic rocks occur at Meillionydd [218 292] and near Crugan Bâch [2125 2995]. At both of these localities the exposures display complicated textures that suggest magma mixing, in which the diorites occur as irregular masses within the more granitic material. Mineralogically, the diorites are essentially the same as those at Carrog but do not have obvious cumulate texture and, consequently, lack a high modal abundance of ilmenite. Clinopyroxene crystals, up to 3 mm in length, may be anhedral in form and may have magnesio-hornblende overgrowths, which form euhedral to subhedral crystals. The extent of alteration of this primary igneous mineralogy varies considerably within samples from the same locality, but, in general, feldspars are sericitised, whilst biotite is chloritised and amphibole transformed to a patchy intergrowth of hornblende and actinolite. Another characteristic of the exposures at Meillionydd and Crugau Bâch is the presence of mafic inclusions of foliated amphibolite, which occur in both dioritic and granitic host rocks. Although commonly extensively retrogressed to an assemblage of chlorite+sericite+ore, less altered samples of these inclusions reveal an amphibolite containing green hornblende+plagioclase+biotite, which define a weak fabric.

Gabbro

Sarn Complex gabbros occur at Graig Fael [217 305]. Exposures here were described by Harker (1888) as 'partially amphibolitised gabbro' and he compared them to the Llangwnnadl outcrops. Matley (1928) described them as 'a plutonic complex of extremely variable texture and composition' that includes granitic rocks and a dolerite dyke, as well as the more typical gabbroic lithologies. Gibbons (1980) emphasised the similarities between the Graig Fael exposures and the dioritic rocks exposed at Carrog and Llangwnnadl, and despite the variations in intensity of foliation interpreted them as igneous rocks rather than metamorphic gneisses. This interpretation was accepted by Beckinsale et al. (1984), who provided the first Rb-Sr isotopic data and suggested an age of 549±19 Ma (an earlier K-Ar age of 317±15 Ma produced by Fitch et al. (1963) on the Sarn Granite is assumed to be a reset age).

The Graig Fael gabbro consists of augite and labradorite in the modal proportion 35:65 and with an overall hypidiomorphic texture. The plagiocase is subhedral, with individual crystals up to 10 mm in length in the coarser grained samples, whilst clinopyroxene ($Ca_{47}Fe_{13}Mg_{40}$) is generally anhedral. Clinopyroxene crystals may be extensively replaced by secondary phases but, where the rock is less altered, are only marginally embayed. In fine-grained samples pyroxene is observed as irregular shaped, monomineralic clots or bimineralic aggregates with interstitial brown titanium magnesio-hornblende. Inclusions or partial rims of the latter around clinopyroxene crystals are thought to be magmatic in origin, having formed as a late-stage magmatic phase, possibly by reaction between clinopyroxene and residual liquid. Where clinopyroxene is extensively altered, there are patchy domains of secondary actinolitic amphibole which may be rimmed by chlorite. Where clinopyroxene has been totally replaced, inclusions and rims of brown hornblende are preserved within the actinolite. Actinolitic amphibole is not considered a magmatic phase. Accessory apatite occurs as inclusions within plagioclase, and chalcopyrite as small blebs within plagioclase and clinopyroxene. Plagioclase is variably altered to sericite, although relatively fresh examples can be found. The coarsest rocks from Llangwnnadl also commonly contain clinopyroxene and are petrologically similar to those from Graig Fael. Brown hornblende is more abundant than at Graig Fael, and subhedral crystals poikilitically enclose or occur interstitially to clinopyroxene. Both these phases define discontinuous mafic banding on a millimetre scale.

Geochemistry

Whole-rock geochemical XRF analyses were undertaken for major and trace elements on representative samples from all major exposures within the Sarn Complex. Rare earth elements (REE) were determined for 18 of these samples by ICP analyses. An additional three XRF major element analyses taken from Beckinsale et al. (1984) were also incorporated into this data set. Selected representatives of these data are given in Table 1.

Silica content ranges from 46 to 79 per cent SiO_2, with gaps at 53 to 61 per cent, 66 to 72 per cent and 75 to 78 per cent. Because of poor exposure, it is not possible to establish whether these represent true compositional gaps or are a product of limited sampling. Fe_2O_3, MgO, MnO, CaO, TiO_2 and P_2O_5 all show negative correlations with SiO_2, and TiO_2 and P_2O_5 show some scatter in the more basic compositions because of varying modal proportions of cumulate apatite and ilmenite. Although Na shows some remobilisation during low-grade alteration, $Na_2O + K_2O$ nevertheless retains a positive correlation with SiO_2. K_2O values increase with fractionation up to 79 per cent SiO_2, this being consistent with an increased modal abundance of potassium feldspar, but they fall to less than 1 per cent K_2O at silica values greater than this. These very high silica–low potassium values are interpreted as resulting from potassium depletion in the most evolved compositions. Contrary to the K-depletion in the Precambrian St David's granophyre described by Blox-

Table 1 Representative XRF analyses for the Sarn Complex.

Wt % oxides	1	2	3	4	5	6	7	8
SiO_2	80.41	79.14	73.47	64.55	61.02	51.26	49.20	52.5
Al_2O_3	12.90	12.92	14.45	17.38	14.38	13.03	15.25	16.8
TiO_2	0.29	0.15	0.49	0.68	0.39	1.60	3.21	0.9
Fe_2O_3	1.56	1.13	3.07	4.62	4.20	11.93	13.80	8.9
MnO	0.01	0.02	0.07	0.10	0.09	0.20	0.25	0.1
MgO	0.22	0.15	0.77	1.94	2.62	7.91	4.77	6.6
CaO	0.11	0.10	0.25	1.64	11.52	8.01	5.92	7.4
Na_2O	3.39	2.64	2.10	3.54	1.26	1.30	2.45	2.4
K_2O	0.79	4.36	3.44	1.75	2.31	1.59	1.25	0.8
P_2O_5	0.02	0.02	0.07	0.21	0.12	1.16	1.45	0.0
LOI	0.71	0.65	1.64	1.75	2.02	1.74	1.85	2.7
Total	100.40	101.28	99.72	98.14	99.92	98.73	99.40	99.5
ppm								
Rb	45	173	120	66	74	48	43	35
Sr	92	61	204	412	79	119	275	275
Ba	408	901	701	625	323	327	276	115
Zr	408	327	247	318	110	124	176	61
Nb	31	27	20	12	10	14	3	3
Y	84	92	39	36	46	42	34	26
Hf	9.6	—	6.7	7.6	—	—	5.8	2.0
Sc	6.3	—	8.8	11.7	—	—	31	35.2
Ta	1.24	—	1.04	0.61	—	—	0.46	0.2
Th	14.5	—	15.1	5.09	—	—	3.21	0.4
U	2.02	—	1.99	0.89	—	—	0.73	0.1
La	61.40	67.90	46.80	33.60	39.2	—	19.20	8.2
Ce	132.20	123.40	99.60	68.80	78.9	—	46.70	19.8
Pr	17.00	19.63	12.75	9.40	10.59	—	8.42	8.2
Nd	60.10	69.60	41.60	32.80	37.70	—	31.60	12.9
Sm	11.50	14.10	8.08	6.25	7.93	—	7.29	3.2
Eu	2.37	1.94	1.61	2.28	1.15	—	2.72	1.0
Gd	10.69	13.35	7.47	6.25	7.95	—	8.98	4.1
Dy	9.81	11.71	5.54	5.29	7.46	—	7.02	4.1
Ho	2.05	2.42	1.23	1.12	1.54	—	1.45	0.8
Er	6.54	7.78	3.94	3.68	4.97	—	4.99	2.9
Yb	5.50	6.63	3.31	3.12	3.97	—	3.19	2.3
Lu	0.84	1.02	0.55	0.50	0.61	—	0.50	0.3

1 Sarn Granite; disused quarry [221 339]
2 Sarn Granite; disused quarry [221 339]
3 Sarn Granite; roadcutting near Pen y gopa [2234 3178]
4 Granodiorite; disused quarry, Carrog Farm [2176 3302]
5 Tonalite; ridge east of Mellionedd Farm [2176 3302]
6 Diorite; Llangwnnadl stream [2100 3293]
7 Diorite; Carrog Farm [2172 3304]
8 Gabbro; quarry, Graig Fael [2175 3040]
LOI = Loss on ignition

ham and Dirk (1988), the Sarn Complex values are considered to be a primary magmatic feature..

Sr v. SiO_2 shows a rather scattered distribution within the basic rocks, which is correlated with a varying plagioclase content or greater alteration of more An-rich plagioclase, whilst the intermediate and acid compositions show a negative distribution consistent with plagioclase fractionation. Plagioclase is also identified as the main phase controlling fractionation in Ba v. Sr and Rb v. Ba plots. Zr values ranging from 61 ppm in the gabbro to 440 ppm in the granite are significantly higher than the values used by Beckinsale et al. (1984) to define high and low Zr groups on their Rb-Sr isochrons. Limited Sm-Nd isotopic data for these rocks (Horák, unpublished data) shows a range of εNd values (calculated at 549 Ma) from -8.2 to -1.4, adding further evidence to the suggestion that $^{87}Sr/^{86}Sr$ within this suite of rocks may not reflect igneous values.

Geochemical discrimination diagrams (Figure 7a, b), which exclude low K samples, emphasise the overall calc-alkaline nature of the Sarn Complex. On an AFM plot (Figure 7a) the basic rocks show considerable scatter and fall mostly within the tholeiitic field. This distribution is attributed to the high FeO^t content in diorites bearing cumulate ilmenite or amphibolitic inclusions rich in TiO_2 and Fe_2O_3. A similar tholeiitic tendency in calcalkaline diorites has been noted by other workers (e.g. Topley et al., 1990; Brown et al., 1990). The affinity of the acid compositions is not well defined on an AFM diagram, but both basic and acid samples fall within the calcalkaline field on a modified Peccerillo and Taylor (1976) K_2O v. SiO_2 plot (Figure 7b), with the more evolved compositions lying in the high-K field. The Rb v. Nb+Y and Rb v. Ta+Yb tectonic discrimination diagrams of Pearce et al. (1984) similarly reflect the volcanic arc affinity of this suite, although the most evolved compositions fall in the within-plate field. This is attributed primarily to the incompatible behaviour of Nb and Y, resulting in their concentration into the most evolved magma compositions.

Normalised rare earth element profiles show a systematic variation with SiO_2 content. All patterns display a tendency towards flat heavy rare earth (HREE) profiles with varying degrees of light rare earth (LREE) enrichment. La_N/Lu_N ratios show little variation within the granite-granodiorite group, suggesting that the observed profile may be an intrinsic feature of the magma, possibly reflecting its source composition, rather than a prod-

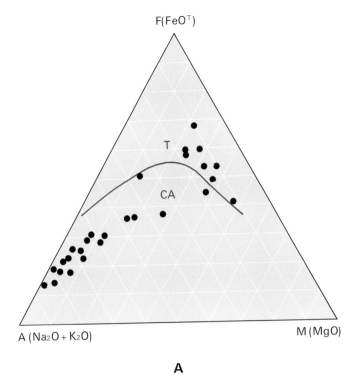

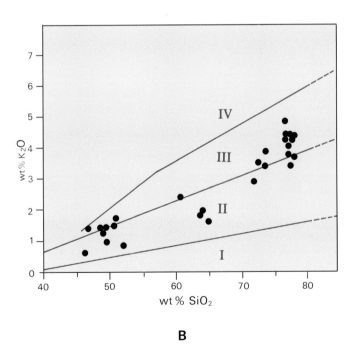

Figure 7 AFM plot (7A) and K_2O v. SiO_2 plot (7B) for plutonic rocks from the Sarn Complex.

uct of fractional crystallisation. An exception to this is the development of a negative Eu anomaly within the granitic rocks, which supports Rb, Sr and Ba data indicating plagioclase fractionation. Similar normalised profiles and contents within the low K and normal K granites suggests that the REE were immobile during low-grade alteration. Dioritic rocks show lower La_N/Lu_N ratios than the granitic compositions, with small positive and negative anomalies reflecting plagioclase accumulation and separation respectively.

PARWYD GNEISSES

The Parwyd Gneisses crop out only over a distance of some 270 m on the mostly inaccessible east face of Parwyd, a precipitous cove cut into the south-west coast of Llŷn [1552 2435 to 1549 2414]. Near the northern end of this clifftop exposure, the gneisses are overlain unconformably by a small outlier of Ordovician sediments. They are bounded to both east and west by fault contacts with Whitlandian (Middle Arenig) sediments.

The Parwyd rocks are coarse, variably foliated orthogneisses, most of which are hornblende-garnet amphibolites, although more felsic varieties are locally exposed. Green hornblende forms up to 80 modal per cent of these amphibolites. Small pale pink garnets up to 2 mm across are also present, most commonly as highly fractured remnants in a groundmass of chlorite. In less retrogressed amphibolites, hornblende may show evidence of alteration to chlorite only along cleavage planes and at the edge of crystals, whilst in the most extensively altered rocks both garnet and hornblende are totally pseudomorphed by chlorite and the rock contains a more pronounced foliation. Accessory phases in the amphibolite include apatite inclusions within hornblende, and a brown prismatic phase (?sphene). Secondary epidote and clinozoisite form stubby, zoned crystals and also cross-cutting veins.

The more felsic lithologies are heavily cataclased and the protolith is uncertain. An assemblage dominated by plagioclase and quartz is generally heavily retrogressed to white mica and chlorite. Accessory phases include prismatic brown crystals of ?sphene/monazite and zircons, 50μm in size. Circular chloritised areas may represent pseudomorphs after garnet, although unretrogressed examples of this phase have not been identified.

The Parwyd Gneisses have previously been placed within the Sarn Complex, although they do not resemble exposures of the latter; they could therefore be quite unrelated. The retrogressive shearing that pervades many collected samples is interpreted as relating to the deformation seen in the nearby Llŷn Shear Zone, although there would presumably have been a greater separation between the Parwyd Gneisses and the mylonitic metasediments on Trwyn Bychestyn (p.23) prior to post-Arenig thrusting. The Parwyd Gneisses have yielded a Rb-Sr whole rock isotopic age of 542±19 Ma (Beckinsale et al., 1984). It remains unclear what this age represents, although the two most likely interpretations are that it records either the amphibolite facies metamorphism or the younger retrogressive shearing.

LLŶN SHEAR ZONE

The Llŷn Shear Zone is a steep zone of semischistose mylonitic rocks that separates the Gwna Mélange from the

Line **X-Y** based on stream section

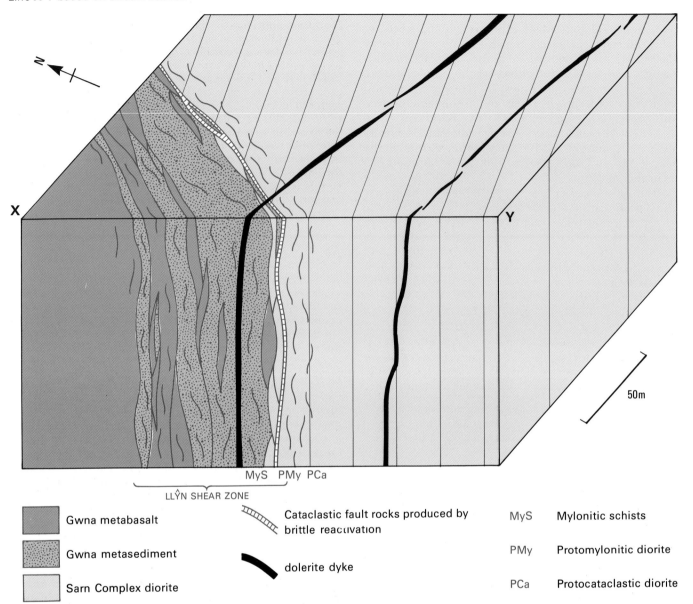

Figure 8 Box diagram illustrating the structure of the Llŷn Shear Zone at Llangwnnadl.

Sarn Complex. It is generally not well exposed, although the Aberdaron district does include one of the most important exposures at Llangwnnadl [2096 3305], where a traverse can be made down the Afon Fawr from dioritic rocks of the Sarn Complex through the shear zone to Gwna basaltic rocks (Gibbons, 1983a). Exposures of typical shear zone lithologies may be seen at the coast around Trwyn Bychestyn [1500 2420]. Other exposures of mylonitic rocks occur on the hill of Mynydd Ystum [1830 2840], along the eastern side of the Afon Daron north-east of Bodwrdda [e.g. 1913 2745; 1922 2766], and at several places between here and Penrhyn Mawr [1900 2618], where they are overlain unconformably by Arenig sediments (Shackleton, 1956).

The rocks within the shear zone are mostly fine-grained, steeply dipping, totally recrystallised, siliceous schists and phyllites. On Trwyn Bychestyn a prograde transition from Gwna Mélange-type sheared metasediments into the schistose shear zone can be traced. In other areas, such as along the south-east side of Mynydd Ystum [1870 2845], the sheared rocks include granites and other lithologies apparently derived from the Sarn Complex. Thus both sides of the shear zone have been affected by the shearing and its margins are gradational.

These mylonitic rocks were correlated by Matley with similar rocks in Anglesey that Greenly had placed within his 'Penmynydd Zone of Metamorphism'. It is now realised that the 'Penmynydd Zone' includes several different schistose rock units of differing grade (e.g. the Anglesey blueschists), and the term is therefore not used in this memoir. The excellent exposures of the siliceous, pale weathering, mylonitic schists on Trwyn Bychestyn [1500 2420] are recommended as the best place to study these distinctive rocks. In addition to this lithology, there are sheared masses of limestone, red phyllite and metabasaltic lava, all of which are presumably derived from the Gwna Mélange. On Trwyn Bychestyn these schists were later thrust over Arenig sediments, removing the eastern margin of the shear zone, although the sheared Parwyd Gneisses occur only a few hundred metres farther east.

Within the Aberdaron area, the Llŷn Shear Zone appears to bifurcate, with one mylonite belt striking south-west through Mynydd Ystum to reach the coast at Trwyn Bychestyn. The other mylonite belt runs south to the coast at Wig. It is not clear whether this bifurcation is an original feature, or whether it is a result of later faulting. The latter is certainly important, as shown by the involvement of the Arenig succession in movements along the Wig and Daron faults (chapter five). If the bifurcation is an early feature of the shear zone, then this may serve to explain the anomalous nature of the Parwyd Gneisses, which are unlike the typical exposures of the Sarn Complex. These gneisses would lie isolated within the shear zone system and need not relate to any other unit exposed elsewhere in North Wales.

Details

Llangwnnadl stream section

Monian rocks are exposed along the banks of the Afon Fawr over a distance of some 600m downstream from Pont Llangwnnadl [2117 3276] (Gibbons, 1983a). Traversing north-west from the bridge, the first exposures are of a coarse-grained, foliated, dioritic rock which, 350 m from the bridge, is in sharp contact with intensely foliated mica schists and compact basic metabasaltic greenschists. Farther downstream, the grade of metamorphism decreases and, in thin section, the rocks display indications of original basic extrusive textures.

The section was first described by Harker (1888) as exposing a 'gneissic diorite' which gave way to a 'massive diorite' near the bridge and became progressively more schistose downstream towards a faulted contact with Monian 'greenschists'. The schistose texture along the contact was attributed to faulting. He correlated the Llangwnnadl diorite with the Graig y Fael exposures of the Sarn Complex. The misidentification of a supposedly intrusive contact between the Sarn Granite and Arenig mudstones at Mountain Cottage Quarry led Harker to assign the Llangwnnadl rocks to the Ordovician. Matley (1928) assigned the Monian schists to Greenly's 'Penmynydd Zone of Metamorphism' and described the dioritic rocks south-east of the contact as 'hornblendic gneisses with a subordinate amount of acid gneiss' which he compared to the Anglesey gneisses. According to this interpretation the Llangwnnadl exposures could not intrude the Sarn Granite (still believed to be Ordovician) and so Matley drew his drift-covered Monian boundary between the two. A more detailed description of the section was published by Shackleton (1956) who referred to Harkers 'gneissic diorite' as 'amphibolite' and 'hornblende migmatites'. By then the pre-Arenig age of the Sarn Granite had been established and Shackleton placed all the 'gneissic' and granitic rocks exposed south-east of the Penmynydd Zone within the same unit (later to be defined as the Sarn Complex by Gibbons, 1980). Shackleton noted that the 'gneisses' were converted to 'crushed green amphibolites' towards the contact with quartz-mica schists and basic greenschists. He noted the presence of igneous textures in the greenschists 140 m north-west of the contact. Whereas Shackleton interpreted this section as displaying a prograde transition from virtually unaltered rocks through schists to 'gneisses', Baker (1969) later argued that the 'gneisses' in fact showed sheared, retrogressive textures against the schists and that the contact was tectonic. Gibbons (1983a) provided a detailed description of the section which illustrated the mylonitic nature of rocks associated with the contact between the 'gneissic diorite' and the schists. The contact was therefore interpreted as a shear zone of unknown (but probably great) displacement. The conclusions of Baker (1969) and Gibbons (1983a) therefore essentially returned to the first interpretation of the contact made by Harker nearly 100 years previously.

The foliated dioritic rocks that crop out in the south-eastern part of this section are described as part of the Sarn Complex on p.18. North-westwards, towards the schistose contact, these foliated dioritic rocks change into a dull greenschist, a transition which becomes noticeable some 18 m across strike from the contact with quartz-mica schists. The relatively coarse texture of the dioritic rocks becomes progressively destroyed by cataclasis (Gibbons, 1983a) towards a sharp, subvertical tectonic contact with fine-grained, quartz-mica schists (essentially metasedimentary mylonites). Immediately downstream from the mylonitic mica schists there are slivers of sheared rocks derived variously from dioritic, metasedimentary and metabasaltic protoliths. These lithologies occur as thin, alternating slivers, intensely sheared within this tectonic contact (Gibbons, 1983a). Particularly good examples of mylonites can be found in the stream bed between the main body of the diorite and the mica schist. Farther downstream, several bands of fine mica schist (becoming fine grained enough to take on more of a phyllitic appearance) weather out above surrounding metabasite and produce a series of small waterfalls. The shear zone at Llangwnnadl is in the order of 100 m wide (Figure 8), thinner than at Trwyn Bychestyn (see below), but comparable to that exposed along strike at Penrhyn Nefyn, 11km to the north-east of the district, where the same contact is exposed at the coast (Gibbons, 1983a).

Trwyn Bychestyn

The cliff section on and west of the headland of Trwyn Bychestyn [1500 2420] provides good exposure from the Gwna Mélange to the finely schistose rocks within the Llŷn Shear Zone. Along the east side of the headland, the schistose rocks are thrust over Arenig sediments (Parwyd Thrust; Gibbons, 1983a), beyond which a steep fault brings the Parwyd Gneisses against the Arenig sediments (Figure 9). The gradation from low-grade Gwna Mélange lithologies into the schistose rocks has been described by Matley (1928), Shackleton (1956), Baker (1969) and Gibbons (1983a).

West of Trwyn Bychestyn, the coast between Porth Felen [1440 2497] and Pared Llech-ymenyn [1463 2465] exposes a series of large, jaspery, basaltic pillow lava clasts associated with a mélange of limestone, jasper, sandstone, 'Gwyddel Beds' and other lithologies, all immersed in a red and green pelitic matrix. The mélange matrix displays a slaty cleavage which dips gently towards the north-west. In places a steeply north-west-dipping cleavage cuts the shallower fabric, this steep cleavage being related to large, south-east-verging upright folds that dominate the sheet dip of this coastal section. In the cliffs at [146 247] a prominent and well-exposed north-westerly dipping thrust zone cuts through the mélange. The thrust zone displays north-west-plunging mineral fibre slickenlines and is associated with a zone of strong north-west-dipping cleavage. The footwall of the thrust shows a brown, bedded, pyritic limestone (Plate 5) resting on a large mass of basaltic pillow lavas. Despite the fact that the thrust lies entirely within the Gwna Mélange, the few metres of rocks caught within the thrust zone are undated cleaved grey mudstones and siltstones, which are more similar to the Ordovician succession cropping out farther east than to anything in the mélange.

At Pared Llech-ymenyn [1463 2465], another north-west-dipping thrust, the Pared Thrust of Gibbons (1983a), occupies the cliff base beneath the Gwna Mélange. In the footwall of this fault there is an ?Ordovician dolerite intrusion cutting steeply dipping, strongly foliated green and red slaty mudstones, sandstones, metabasalt, white quartzite and brown limestone. These steeply dipping rocks, interpreted as highly deformed Gwna Mélange, become increasingly deformed and recrystallised towards the south-east. A large white mass referred to by Matley (1928) as the 'Great Quartzite' forms a conspicuous landmark on the cliffside. About 130m north-west of this quartzite, a disrupted and highly deformed mixture of grey-green and purple phyllitic metasediments and basic lavas contains irregular masses of purple and dark green jaspery metasediments.

Most of the headland of Trwyn Bychestyn consists of sheared and recrystallised, pale weathering metasediments, locally containing lenses of deformed limestone and, on the extreme southern tip of the headland, metabasalt. An overall steep dip is maintained across the headland, with minor upright folds (related to the larger south-east-verging folds mentioned earlier) being modified by flatter-lying south-east-verging minor folds interpreted as being related to the underlying Parwyd Thrust (Plate 7). On the west side of the headland, there is a subtle but perceptible gradation from phyllitic rocks into fine schists, with white mica crystals becoming visible to the naked eye and forming aggregates that reach up to 4 mm in length.

Mynydd Ystum

Fine-grained, mylonitic schists form the prominent hill of Mynydd Ystum [1870 2845] that rises to a height of some 150 m above sea level, 2 km north-north-east of Aberdaron. The locality was originally mentioned by Blake (1888) who recorded the presence of a 'terribly squeezed and broken' granite on the east side of the hill and stated that the main mass of the hill consisted of 'dark foliated mica schists'. These rocks were later mapped by Matley (1928) as typical 'Penmynydd Mica-schists'. Matley also noted the presence of deformed coarse-grained rocks on the east side of the hill and referred to these as acid gneiss and 'tremendously crushed' basic gneiss. Shackleton (1956) interpreted the exposures in terms of his prograde transition hypothesis, whereas Baker (1969) and Gibbons (1983a) both recognised mylonitic textures and interpreted the schist belt as a shear zone lying between the Sarn Complex and unexposed Gwna Mélange to the north-west. Mylonitised granitic rocks can still be seen today [188 297], although such representatives of the Sarn Complex are generally very poorly exposed. The main body of the hill exposes fine-grained, sheared metasediments similar to those exposed on Trwyn Bychestyn. Locally, exposures reveal upright, south-west-plunging folds [e.g. 1897 2863] similar to those seen on Trwyn Bychestyn. The lack of flat-lying folds on Mynydd Ystum may be attributed to the lack of late thrusting at the schist-Sarn Complex contact.

Pont Cyll-y-Felin

One kilometre west of Mynydd Ystum, a series of exposures emerge from the drift cover. These exposures are all of metabasalts belonging to the Gwna Mélange, and are interpreted to be part of the same sequence seen farther west on Mynydd Carreg [1640 2917]. The most extensive of these exposures is to be found along the banks of the Afon Cyll-y-Felin [1750 2835]. To the north [1749 2857] and south [1745 2822] of the stream there are exposures of undeformed jaspery pillow lavas. Within the stream itself, however, the basic volcanic rocks are strongly deformed and possess a subvertical east–west-striking foliation (parallel to that on Mynydd Ystum) and a fine-grained greenschist facies mineralogy. This shear zone is some 10 m wide and is interpreted as a splay from the main schist zone to the south and east.

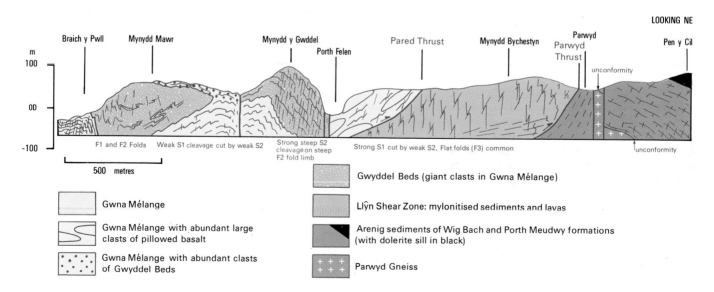

Figure 9 Sketch structural cross-section across the south-west coast of Llŷn [approximately 135 259 to 159 240].

Plate 7 Steeply south-easterly dipping mylonitic metasediments within the Llŷn Shear Zone on Mynydd Bychestyn [1497 2430]. The finely schistose mylonitic fabric is folded by both upright, south-easterly verging structures (centre) and by later folds with a flat-lying axial surface (dipping gently to the left). Looking north-east.

Afon Daron, Penrhyn Mawr and Wig

One kilometre south-east of Mynydd Ystum, exposures of solid rock along the narrow gorge of the Afon Daron reveal Ordovician mudstones and sandstones that are abruptly terminated to the east by the Daron Fault, beyond which sheared Sarn Complex rocks crop out [1923 2770; 1913 2747]. These Sarn Complex rocks are thoroughly mylonitised and display a subvertical shear zone fabric. 450 m to the south, another exposure of Sarn Complex rocks [1916 2698] displays traces of a granitic texture. Beneath the Arenig unconformity at Penrhyn Mawr [1902 2619], similarly sheared granitoid rocks are exposed. Sheared representatives of the Sarn Complex and Gwna Mélange are exposed in several small disused quarries farther south-west towards the coast south of Pen-Cae [1891 2669]. On the coast at Wig, 500 m south-west of Penrhyn Mawr, subvertical, sheared Monian rocks are exposed, faulted against (and locally lying unconformably beneath) Arenig sediments. The mylonitic rocks here are of uncertain protolith, although their general appearance, with deformed lenses of quartzitic material surrounded by a semipelitic matrix, suggests derivation from the Gwna Mélange.

TECTONIC SETTING

The oldest rocks exposed in the district comprise two primary tectonostratigraphic units: a mélange (Gwna Mélange) and a calc-alkaline predominantly plutonic complex (Sarn Complex). These two units are juxtaposed by a steep mylonite belt (Llŷn Shear Zone). The Gwna Mélange displays a highly disrupted lithostratigraphy that includes strata produced in a variety of settings. The semipelitic matrix and grey-green sandstones and siltstones that form the bulk the mélange are interpreted as disrupted marine strata originally deposited within an active plate margin setting such as a trench or forearc basin. The abundant jaspery pillow basalts are commonly erupted into limestones and show similarities with ocean island sequences. In contrast, many of the other clast assemblages, particularly those common in the upper part of the Gwna Mélange, suggest shallow water or even continental conditions (e.g. white orthoquartzite, limestones) and a connection with continental basement (granitic clasts). Given the likelihood of an active plate margin setting for these rocks, a calc-alkaline signature to the Sarn Complex and the occurrence of blueschists farther north-east along strike, it is likely that the Gwna Mélange was produced by a process directly associated with oceanic plate subduction or plate collision. The wide mixture of depositional environments recorded by the mélange clasts suggests a mixing of oceanic and continental materials. Although an accretionary prism setting is possible, it has been suggested that a collision between an arc and a continental margin could more likely produce the extraordinary heterogeneity typical of the Gwna Mélange (Gibbons, 1983b). A modern analogue for this scenario is provided by the recent collision of Taiwan with continental China, in which continental margin, accretionary prism and forearc basin rocks were disrupted and mixed to produce a regional-scale mélange (Lichi Mélange of Page and Suppe, 1981).

The only potential connection between the Gwna Mélange and the Sarn Complex is provided by the granitic clasts in the mélange on Bardsey Island, which were possibly derived from the Sarn Complex. Given the great contrast in geology on each side of the Llŷn Shear Zone, and the uncertainty over any correlation between the Gwna Mélange and the Sarn Complex, each of the latter two units may be considered as suspect terranes, i.e. fault-bounded geological entities of regional extent, each characterised by a geological history that is different from the neighbouring terrane (cf. Coney et al., 1980; McWilliams and Howell, 1982; Gibbons, 1989b). Implicit in this conclusion is the suspicion that each of these suspect terranes could well be highly displaced with respect to each other and/or the nearest cratonic basement existing in Late Precambrian times.

The tectonic setting of these rocks is therefore interpreted to have been an active plate margin with calc-alkaline magmatism producing a wide range of late Precambrian ('Avalonian' and 'Cadomian') rocks. The Gwna Mélange represents a disrupted sequence of sediments that were probably mostly deposited along the active plate margin, but which also includes rocks of possible 'exotic' oceanic origin. Following the disruption event that produced the Gwna Mélange, this latter unit was faulted against the plutonic roots of an arc (Sarn Complex). Steep ductile faults such as the Llŷn Shear Zone are found throughout the Avalonian/Cadomian area and have been interpreted as having been produced during a phase of dominantly transcurrent suspect terrane dispersion along the active plate margin during oblique convergence (Gibbons, 1990b). Following, or during this phase of terrane dispersion, subsidence to the south-east produced basins in which Cambrian sediments now exposed at St Tudwals and Harlech were deposited.

THREE
Ordovician rocks

The Ordovician rocks in the district were originally deposited as mostly sands, silts and muds on the sea floor, following a marine transgression over an uplifted and eroded basement. The oldest Ordovician rocks, where exposed, rest with marked angular unconformity on the rocks of the Llŷn Shear Zone and on the Parwyd Gneisses. The Ordovician rocks crop out in two main areas separated by the Daron-Wig fault system (Figure 1). The westerly area includes the district around Aberdaron and the western side of Aberdaron Bay; the other is in the south-eastern corner of the district, eastwards from the east side of Aberdaron Bay. Exposure is poor inland, except for a few localities such as the Afon Daron gorge, but excellent at the coast (Plate 9).

ABERDARON BAY GROUP

All the Ordovician (Arenig to Llanvirn) rocks have been placed within the Aberdaron Bay Group; they comprise dark grey mudstone and laminated siltstone, with subordinate fine-grained, grey sandstone and conglomerate. Ironstones occur at two distinct levels. The group is subdivided, in ascending order, into the Wig Bâch Formation (Moridunian to Fennian age), the Porth Meudwy Formation (Fennian) and undivided beds above that, which range up into the Llanvirn (Figure 10). The beds on both sides of the Daron Fault dip generally south-eastwards so that the youngest rocks (Llanvirn) are found both immediately west of the Daron Fault and in the

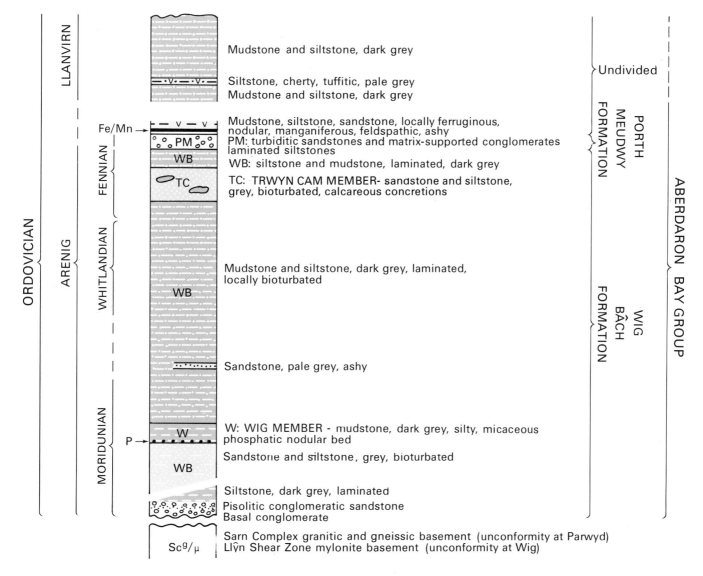

Figure 10 Composite lithological log of the Aberdaron Bay Group in the Aberdaron Bay area. Scale: 1 cm to 50 m.

extreme south-east corner of the district. The recognition of the Moridunian, Whitlandian and Fennian stages is based upon palaeontological work by Beckly (1985, 1988) and additional data collected during this survey. So far, Moridunian strata have been identified east of the Daron Fault only, although the small exposure of Ordovician sediments resting on the unconformity at Parwyd may be of this age. The Ordovician sedimentary succession is estimated to be over 700 m thick.

Initial mapping of the Ordovician outcrop was achieved by Matley (1928, 1932), who recognised both Arenig and Llanvirn rocks within the district. Detailed biostratigraphical work by Beckly (1985, 1987, 1988) led to a lithostratigraphical subdivision of the Arenig rocks into three formations ('Sarn Formation', 'Aberdaron Formation', 'Carw Formation'). His 'Sarn Formation' basal stratotype, defined at the poorly exposed Mountain Cottage Quarry [2300 3470], outside the district, was interpreted as being of Fennian age. However, the best exposure of Beckly's 'Sarn Formation' was taken to be that on the coast at Porth Cadlan [2000 2601], where it was interpreted as being of Moridunian age and equivalent to exposures further west at Wig Bâch [1861 2572]. The 'Aberdaron Formation' of Beckly (1988) includes most of the Arenig exposures on both sides of Aberdaron Bay and spans all three Arenig stages, from Moridunian to Fennian. He assigned the overlying sequence of more volcaniclastic sediment to the 'Carw Formation' (type area at Nant-y-Carw [2352 3233 to 2355 3232] outside the district). However, within the Aberdaron area, the base of the proposed 'Carw Formation' was not clearly recognisable, and none of the boundaries between the three formations could be traced inland from the coast. It has therefore been necessary to introduce a new lithostratigraphy, using clearly defined, mappable formations. In areas of poor exposure, such as inland north of Aberdaron and in the south-east corner of the district, the Ordovician rocks are mostly defined from the adjacent region.

Plate 8 View looking west across the coastal exposure of the Aberdaron Bay Group from the summit of Mynydd Penarfynydd. Boulders of cariously weathered picrite (left foreground) are from the underlying layered intrusion. The coast in the middle ground shows south-easterly dipping Ordovician sediments (Wig Bâch Formation) and dolerite with the tidal islet of Maen Gwenonwy lying just right of centre. Beyond the small islands of Ynys Gwylan-fawr and Ynys Gwylan-bâch (left) rises the hill (Mynydd Enlli) that forms the north-eastern part of Bardsey Island.

West of the Daron Fault, the basal part of the Wig Bâch Formation and the overlying Porth Meudwy Formation are present. The former corresponds broadly to the 'Sarn Formation' and part of the 'Aberdaron Formation' of Beckly (1988 — sub-area 1). The Porth Meudwy Formation, the most distinctive lithostratigraphical unit in the Aberdaron district, corresponds to the Porth Meudwy Member of Beckly (1988 — sub-area 1). The rocks lying above the Porth Meudwy Formation (corresponding to part of the 'Aberdaron Formation', and the 'Carw Formation' of Beckly 1988 — sub-area 1) have not been given formational status but are placed within the Aberdaron Bay Group (undivided). The highest part of the Aberdaron Bay Group west of the Daron Fault is well exposed along the west side of Aberdaron Bay where it is overlain by a thick dolerite sill, above which there is no exposure.

East of the Daron Fault, the Ordovician sequence is exposed at the coast between Wig Bach and Porth Ysgo (Plate 8), and belongs to the Wig Bâch Formation i.e. the lowest part of the succession. This area provides the only good exposure of the lower part of the Wig Bâch Formation which, west of the Daron Fault, has been removed by faulting. Farther east from Porth Ysgo [2075 2640], the Ordovician sequence includes both Arenig and Llanvirn rocks, but structural complications combined with extremely poor exposure have not permitted subdivision of the upper part of the sequence, which has been depicted on the map as Aberdaron Bay Group, undivided.

Palaeontology

The dark grey mudstones and laminated siltstones that typify most of the Aberdaron Bay Group were deposited in relatively quiet water and contain a marine fauna of trilobites (Plate 9), graptolites (Figure 11) and brachiopods. Although fossils are present at several localities, they are only locally abundant. Localities that have yielded abundant and/or varied Arenig faunas include the old quarry at Dwyrhos [1672 2649], the cliffs above Ogof Lleuddad [1903 2561], the north side of Maen Gwenonwy [2006 2599] and several places in Nant y Gadwen [2110 2663]. Llanvirn fossils occur at the coast beneath the Mynydd Penarfynydd layered intrusion and behind the farmhouse at Penarfynydd Farm [2189 2663], where dark grey mudstones and siltstones interbedded with pale tuffaceous horizons contain a rich fauna of trilobites and graptolites.

The graptolites found in the district are most readily correlated with the graptolitic succession in the English Lake District (Jackson, 1962), whereas the shelly faunas may be correlated with the stage and zonal divisions proposed for South Wales by Fortey and Owens (1987) and shown in Table 2. Correlation between these two zonal schemes is good at the base of the Llanvirn and the Fennian but is uncertain at the level of the Whitlandian and increasingly insecure lower in the succession.

Many of the fossils from the district were listed by Beckly (1988, p.325) and the present account is based largely on his work, together with Matley's collection and some new material now in the collection of the British Geological Survey. Biostratigraphical subdivision of the succession can be attempted (following Beckly, 1988), even though many of the fossils are poorly preserved, and the results based on them are accordingly tentative.

Moridunian Stage

The principal evidence for Moridunian strata lies in the occurrence in the Wig Bâch Formation of *Merlinia selwynii* (Salter), the zonal species for the lower Moridunian in South Wales. This species occurs in dark mudstones at Maen Gwenonwy [2004 2598], where it appears to be represented by the 'late form' of the species, transitional to *M. rhyakos* (Fortey and Owens, 1978); it is associated there with *Azygograptus eivionicus* Elles (Figure 11), apparently an early form of the species (Beckly, 1988, p.332). Beds of similar lithology at Wig [1865 2568], 4 to 5 m above the phosphatic 'junction bed' of Beckly (1988), yielded a similar association, but the fossils there are too poorly preserved to be certain that the species are the same as those at Maen Gwenonwy. Numerous distorted *Merlinia* were also collected from mudstones at Ogof Ddeuddrws [1868 2550].

Well-preserved *Merlinia selwynii* have been collected from the structurally complicated area at the northern end of Nant y Gadwen, and these are provisionally taken to indicate the presence of Moridunian strata there. At Benallt Mine, however, *M. selwynii* is accompanied by a specimen of *Hanchungolithus* cf. *primitivus* (Born) (Plate 9) and *Furcalithus*? sp. These, though not definitive, suggest the possibility of correlation with the 'mid-Arenig' faunas abroad and with the Whitlandian Stage (Beckly, 1988, p.332), implying that *M. selwynii* may have a longer range in North Wales than in South Wales.

Dr S G Molyneux reports that an acritarch assemblage from beds at Maen Gwenonwy includes *Uncinisphaera*? spp. D and E of Molyneux (1987), both of which occur in Moridunian strata in South Wales.

Whitlandian Stage

Several trilobite species typical of the Whitlandian Stage occur in the district, *Cyclopyge grandis grandis* (Salter) (Plate 9) being the most widespread. At Dwyrhos Quarry [1672 2649] it is accompanied by a rich fauna including *Segmentagnostus hirundo* (Salter), *Cnemidopyge salteri* (Hicks), *Shumardia gadwensis* Fortey and Owens, *Bohemopyge scutatrix* (Salter), *Furcalithus* aff. *sedgwickii* (Salter) and a species of *Psilacella* (Plate 9). A similar fauna, but lacking the first two species, is known from the eastern side of Nant y Gadwen [2119 2679]. Beckly collected *C. grandis grandis* and the zonal species *Gymnostomix gibbsi* from the faulted outcrop of Parwyd Bay [1543 2439]. None of these outcrops is in good stratigraphical continuity with strata referred to the Moridunian Stage. The beds west of Ogof Lleuddad [1901 2559], which are interpreted as overlying the Moridunian of Wig and Maen Gwenonwy, yielded less satisfactory evidence of the Whitlandian, namely unidentified species of *Furcalithus* and *Novakella*?, and a *Cyclopyge* akin to *C. grandis* but differing in the length of the py-

30 THREE ORDOVICIAN ROCKS

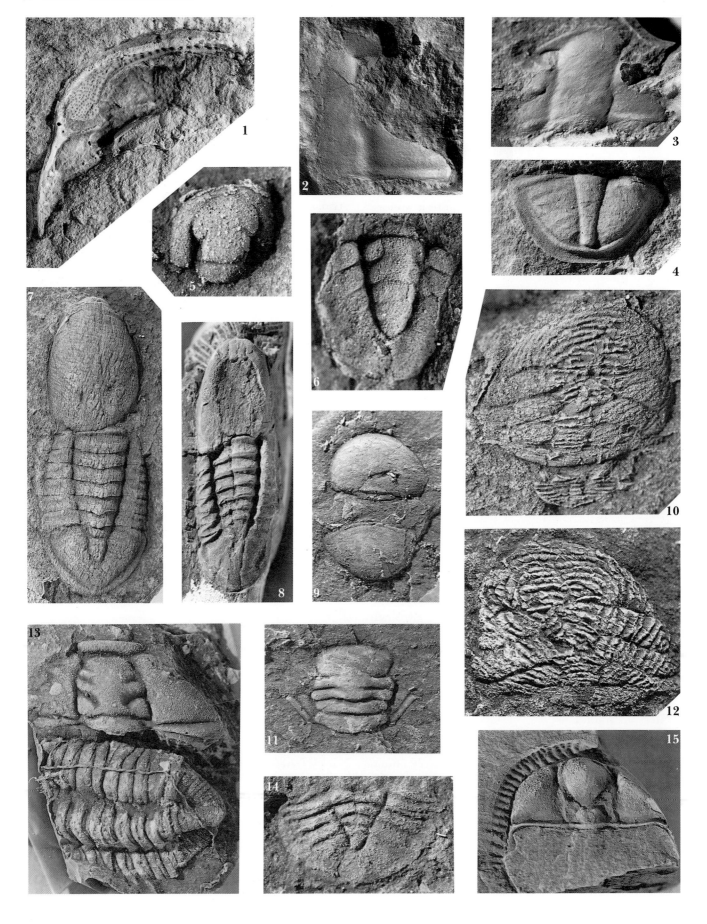

Plate 9 Examples of Ordovician trilobites from the district. Specimen numbers with the prefix It are from the A J Beckly collection in the British Museum of Natural History. All specimens whitened. Photographs by Philip Wells.

1. *Hanchungolithus* cf. *primitivus* (Born). Ventral view of lower lamella. It 19861a (latex cast), × 10. Arenig, Moridunian. Just west of Clip y Gylfinhir.

2–4. *Merlinia selwynii* (Salter). The cranidium in Figure 2 shows terrace-ridges on the front of the glabella. 2 and 4 are BGS Z4131 and Z4122 (Matley coll.), from shaft and adit west of Clip y Gylfinhir. 3 is It 21220, from north end of Nant y Gadwen. All Arenig, Mordunian, all × 2.5.

5,6. *Shumardia* sp. Cranidium and pygidium. It 2122b, 21227b (latex casts), × 15. Lower Llanvirn, Penarfynydd.

7. *Cyclopyge grandis grandis* (Salter). It 21222b (latex), × 5. Arenig, Whitlandian, Dwyrhos Quarry.

8. *Cyclopyge* aff. *grandis* (Salter). Compressed specimen showing proportionally wider pygidial doublure than *C. grandis grandis*. BGS RX 2775, × 5. Arenig, Whitlandian (?), Ogof Lleuddad.

9. *Microparia* sp. Cranidium and pygidium whose transverse shape recalls *Microparia* sp. B of Kennedy, 1989. It 21224b (latex), × 5. Lower Llanvirn, Penarfynydd.

10,12,14. *Psilacella* sp. nov. Two cranidia showing long glabellar furrows and distinctive sculpture. It 19866, 19867 (latex casts). Associated pygidium. It 19868, All × 10, all Arenig, Whitlandian, Dwyrhos Quarry.

11. *Bohemilla* (*Fenniops*) sp. Cranidium having the shape of *B.* (*F.*) *kloucekі* Marek, but the counterpart has an anterior border like that of *B.* (*F.*) *sabulon* Fortey and Owens. It 21225b (latex), × 5. Lower Llanvirn, Penarfynydd.

13. *Platycalymene tasgarensis tasgarensis* Shirley. It 21223a (latex), × 2.5. Lower Llanvirn, Penarfynydd.

15. *Stapeleyella murchisoni* (Salter) Cranidium showing a few Y-shaped ridges on the fringe anterolaterally. National Museum of Wales 27.110.G271 (G J Williams, coll.), × 5. Lower Llanvirn, Penarfyndd.

gidial axis and the width of the pygidial doublure (Plate 9). Dr Molyneux reported a poor, undiagnostic acritarch flora from rocks at Ogof Lleuddad.

Fennian Stage

Rocks of Fennian age have been proved at Nant y Gadwen and Benallt, but have not been proved satisfactorily at other localities in the district. Beckly collected several species of trilobites, especially from Nant y Gadwen, but of these only the cyclopygids *Cyclopyge grandis brevirhachis* Fortey and Owens, *Microparia broeggeri* (Holub) and *Pricyclopyge binodosa eurycephala* Fortey and Owens can be used to correlate with the Fennian of South Wales.

Graptolites from the west side of Nant y Gadwen [2108 2661] include *Didymograptus* cf. *extensus linearis* Monsen, *D. uniformis lepidus* Ni, *D. distinctus* Harris and Thomas (Fortey et al., 1990), *Isograptus caduceus gibberulus* (Nicholson) (Figure 11), *Pseudisograptus angel* Jenkins and *P.* cf. *dumosus* (Harris) (Fortey et al., 1990); all these occur in the *gibberulus* Biozone in the Lake District, except the last named which is not known there. *Didymograptus hirundo* (Salter) (Figure 11), which occurs in both the *gibberulus* and *hirundo* biozones in the Lake District, is well represented at Nant y Gadwen (including one specimen, BGS AW 27, originally 80 cm long). At the northern end of Nant y Gadwen, Dr Beckly collected *D.* cf. *extensus linearis*, *D. distinctus* and *P.* cf. *dumosus*. The association of *D. distinctus* and *P.* cf. *dumosus* with a *gibberulus* Zone fauna is valuable as affording a correlation with the upper Castlemainian (Ca3) of the Australasian succession.

The age of the Porth Meudwy Formation is not well constrained. *Tetragraptus reclinatus* Elles and Wood, which has been collected from underlying strata, has a long stratigraphical range. Beckly (1988, p.325) tentatively assigned fossils from above the formation (his Locality 12) to the top of the Fennian, but they are here considered

Table 2 Chrono- and biostratigraphical classification of the Aberdaron Bay Group.

	South Wales		Lake District	Ordovician lithostratigraphy (Position of Porth Meudwy Formation is approximate)	
Series	Stage	Biozone	Biozone		
Llanvirn	—	*Didymograptus artus*	*Didymograptus artus*	Undivided	
Arenig	Fennian	*Dionide levigena* *Bergamia rushtoni* *Stapeleyella abyfrons*	*Didymograptus hirundo* ——————— *Isograptus gibberulus*	— ? — ? — ? — ? — ? — ? Porth Meudwy Formation ———————	Aberdaron Bay Group
	Whitlandian	*Gymnostomix gibbsii* *Furcalithus radix*	*Didymograptus nitidus* ———————	Wig Bach Formation	
	Moridunian	*Merlinia rhyakos* *Merlinia selwynii*	*Didymograptus deflexus*		

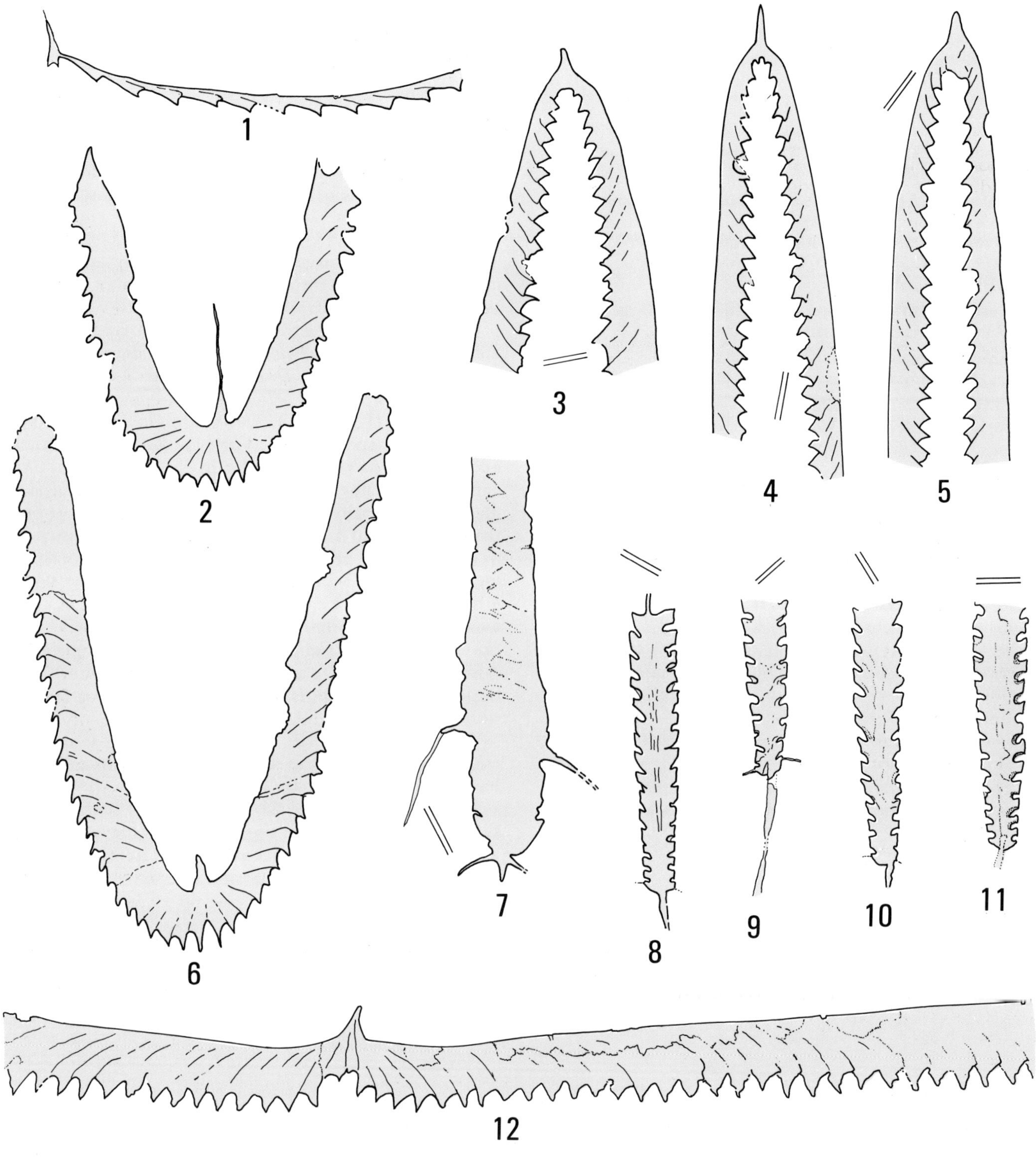

Figure 11 Examples of Ordovician graptolites from the district. Specimens numbered with the prefix Q are from the A J Beckly collection in the British Museum of Natural History. The parallel lines indicate the direction of tectonic lineation. All magnified ×5.

1. *Azygograptus eivionicus* Elles. Q5872a. Arenig, Moridunian. Maen Gwenonwy.
2,6. *Isograptus caduceus gibberulus* (Nicholson). Both on BGS AW24. Arenig, Fennian. Nant y Gadwen.
3–5. *Didymograptus* (D.) *spinulosus* Perner. BGS RX 3824, National Museum of Wales 27.110.G776 (G J Williams coll.), BGS RX 3825. Lower Llanvirn, Penarfynydd.
7. *Lonchograptus* sp. The spines above the base are much thinner than those of L. *ovatus* Tullberg. Q6356. Lower Llanvirn, Penarfynydd.
8–11. *Amplexograptus confertus* (Lapworth), showing virgellar prolongation. BGS RX 3827, RX 3826, Q6360a, Q6357. Lower Llanvirn, Penarfynydd.
12. *Didymograptus* (s.l.) *hirundo* Salter. Q5707b. Arenig, Fennian. Nant y Gadwen.

more likely to be referable to the Llanvirn Series (see below).

Llanvirn Series

The Llanvirn Series is recognised by comparison with the succession in South Wales, where the base is drawn at the appearance of pendent *Didymograptus* such as *D. (D.) artus* Elles and Wood and *D. (D.) spinulosus* Perner (Fortey et al., 1990), though several other species range across this boundary.

A large lower Llanvirn assemblage has been collected at Penarfynydd [2189 2663]. It includes the brachiopods *Lingulella?* and *Paterula*; trilobites (Plate 9) such as *Barrandia homfrayi* Hicks, *Bergamia?* sp., *Bohemilla (Fenniops)* sp., *Dindymene* cf. *didymograpti* (Whittard), *Dionide levigena* Fortey & Owens, *D. turnbulli* Whittington, *Microparia* sp. B of Kennedy, 1989, *Platycalymene tasgarensis tasgarensis* Shirley, *Shumardia* sp. and *Stapeleyella murchisoni* (Salter); graptolites (Figure 11) including *Amplexograptus confertus* (Lapworth), *Cryptograptus* sp., *Didymograptus (D.) spinulosus, D. (D.) geminus* (Hisinger), *Diplograptus ellesi* Bulman, *Eoglyptograptus* sp. and *Lonchograptus* sp. The graptolites are indicative of the lower Llanvirn Zone of *Didymograptus artus* (formerly known as the '*D. bifidus*' Zone). Many of the trilobites are most similar to species that range across the Arenig–Llanvirn boundary in South Wales (Fortey and Owens, 1987, p.92), but three, *D. turnbulli, P. tasgarensis tasgarensis* and *S. murchisoni*, appear to be restricted to the lower Llanvirn.

From the Afon Daron [1897 2740], Beckly (1988, p.328) collected *Platycalymene* sp. and indeterminate pendent and biserial graptolites. He considered these to be strongly indicative of a Llanvirn age. More debatable is the significance of the assemblage from above the Porth Meudwy Formation at Porth Meudwy [1645 2559]. Beckly collected *Didymograptus* aff. *nicholsoni* Lapworth, *Cryptograptus tricornis schaeferi* Lapworth and *Pseudoclimacograptus* sp. From the same place Dr T Young collected slender *Janograptus*-like specimens, *Diplograptus* cf. *ellesi* Bulman and some *Eoglyptograptus?* with closely set thecae. None of these taxa is known with certainty from the Arenig but because *D. ellesi* and *C. tricornis schaeferi* are recorded from the Llanvirn, the locality is here tentatively referred to the Llanvirn Series. The graptolites do not, however, resemble those in the more typical lower Llanvirn assemblage at Penarfynydd.

Wig Bâch Formation

The sediments of the Wig Bâch Formation, up to 550 m thick, were deposited during the period of approximately 10 million years represented by the Arenig Series (Moridunian to Fennian) and comprise dark grey mudstone, laminated siltstone and bioturbated sandstone. Typical dark mudstones and siltstones are well exposed at the coast between Porth Simdde [1666 2622] to Trwyn Cam [1665 2603], and across the headland on the east side of Aberdaron Bay as far as Ogof Lleuddad [1908 2552]. Deposition of these dark muddy rocks was interrupted by periodic influxes of sand into the area, most notably during the deposition of prominent sandstone units such as those exposed at Porth Ysgo [2075 2650] and at the back of Porth Meudwy [1633 2551]. There are two main sandstone sequences, the lower of which occurs above the unconformity at Wig Bâch, Porth Cadlan and Porth Ysgo, and is Moridunian in age. The higher sandstone crops out between Trwyn Cam and Porth Meudwy, to the west of Aberdaron, and possibly on Ebolion to the east of Aberdaron, and is Fennian in age. These sandstones are commonly full of the traces of burrowing organisms, and highly bioturbated surfaces are exposed at several coastal localities, such as in Porth Meudwy [1633 2551]. In some places the sediments contain large diagenetic calcareous concretions, good examples of which occur at the back of Parwyd [1545 2440], north of Trwyn Cam [1665 2605] and on Ebolion [1888 2514].

The base of the Wig Bâch Formation crops out at Wig [1863 2570], Parwyd [1549 2434] and Penrhyn Mawr [1902 2616]. The first two localities were described by Matley (1928, 1932) and the third by Shackleton (1956). Unfortunately, the contact at Penrhyn Mawr has been obscured by concrete, and the Parwyd outlier is perched high on the cliffline in a position extremely dangerous of access. The remaining exposure, on the coast at Wig, although very restricted in extent and also difficult of access, clearly displays approximately 3 m of south-east-dipping clastic sediments resting unconformably on vertical mylonites (Gibbons, 1989a). Above a basal 30 cm-thick breccio-conglomerate are six beds of brown ferruginous sandstone, varying in thickness from 4 to 20 cm and interbedded with thin mudstones. These are capped by a thin (13 cm) conglomeratic sandstone containing ferruginous ('chamositic') ooids, most of which are phosphatised, and fragments of inarticulate brachiopods (T P Young, personal communication). The phosphatic and ferruginous sediments are interpreted as having been formed as a condensed sequence during sediment starvation on the drowned shelf. Above the winnowed phosphatic conglomeratic sandstone is an abrupt change to dark grey mudstones with laminations of grey siltstone and fine sandstone. The unconformity surface dips at 55° towards the ESE and terminates against north–south faults to both west (against Monian mylonites) and east (against the Wig Fault).

On the coast at Wig Bâch [1860 2567], immediately east of the Wig Fault, Ordovician sediments are well exposed, crumpled against the fault plane. Thinly bedded grey sandstone and siltstone coarsen upwards into highly bioturbated grey sandstones with both vertical and horizontal burrow structures such as *Phycodes* (Crimes, 1969). Lithologically, the highest unit exposed in the sediments west of the Wig Fault are similar to the lowest beds to the east of the fault, and so there may not be a large post-Arenig displacement across this fault. At the top of the bioturbated sandstones at Wig Bâch is a distinctive bed of pebbly, bioturbated, phosphatic sandstone, 0.28 m thick (the 'junction bed' of Beckly (1988). This indurated bed is interpreted as a hardground produced during a period of nondeposition.

Sandstones of the Wig Bâch Formation are also exposed above a thick dolerite intrusion at Porth Ysgo [2070 2648] (Plate 15) and at Porth Cadlan [2003 2518].

The sandstones are much bioturbated, well sorted, and pale grey in colour due to bleaching by the intrusion. The sandstones are interbedded with siltstones and are locally cross-stratified and pebbly, with thin mudstone intercalations. The sandstones locally contain phosphatic material, especially concentrated along erosive surfaces. Dominantly horizontal burrows are common trace fossils, produced by deposit feeders such as *Phycodes* and *Teichichnus* (Crimes, 1969, 1970). This 'flaggy sandstone facies' (Beckly, 1985, 1988) is similar to that beneath the 'junction bed' exposed at Wig Bâch; Beckly correlated these exposures and interpreted the sandstones at Porth Cadlan to represent the body stratotype of his Sarn Formation. The nodular 'junction bed' is unexposed at Porth Ysgo and Porth Cadlan.

Wig Member

The 'junction bed' at Wig Bâch is overlain by about 30 m of monotonous dark silty grey mudstone and grey siltstone with phosphatic nodules, which constitute the Wig Member of Beckly (1988). Numerous distorted fragments of *Merlinia* have been collected from mudstones at Ogof Ddeuddrws [1868 2550], and Beckly (1988, p.332) records *Merlinia selwynii* and *Azygograptus* fragments from the Wig Member at Wig Bâch, which he interprets as probably of Moridunian age. The dark mudstone facies typical of the Wig Member is also well exposed on the tidal island of Maen Gwenonwy [2007 2599] where a Moridunian faunal assemblage that includes *Merlinia selwynii*, *Azygograptus eivionicus* and *Didymograptus praenuntius* has been recorded (Beckly, 1988). *Merlinia* collected from Maen Gwenonwy during the survey proved to be the 'late form' of the species, transitional to *M. rhyakos*, and is associated with an apparently early form of *Azygograptus eivionicus*.

The top of the Wig Member grades up into several hundred metres of dark mudstone, siltstone and fine-grained sandstone, with rare ashy sandstone beds. These beds crop out around the precipitous headland of Trwyn y Penrhyn, where they generally dip south-east away from the Wig Fault and become more strongly cleaved and faulted closer to that fault. One locality within this sequence, on the clifftop west of Ogof Lleuddad [1902 2560], was interpreted as Whitlandian in age by Beckly (1988). Further study of this area yielded less satisfactory evidence for a Whitlandian age (*Furcalithus*, *Novakella*? and a *Cyclopyge* akin to *C. grandis*) as outlined on p.30. Previously, the Maen Gwenonwy exposures were considered to lie structurally above the apparently younger (Whitlandian) rocks exposed further south-west at Ogof Lleuddad, and their position was explained as due to repetition by faulting (Beckly, 1988 p.332). However, the relevant fault, exposed at the back of a cave (Ogof Lleuddad) [1907 2553], proved to be only a small, splaying fracture with identical sediments on either side. More significant is that, immediately to the east of the fault, the contact between Matley's Gallt-y-Mor 'Sill' (Matley, 1932) and the sediments shows the dolerite intrusion to be highly transgressive and dyke-like in form. Given the cross-cutting nature of the dolerite intrusion, extrapolation of the sequence north-eastwards from Wig Bâch to Porth Cadlan presents no problem. The sandstones are cut out at the base of the intrusion, but continue further east above the intrusion at Porth Cadlan (where the top of the intrusion has a sill-like form).

Trwyn Cam Member and overlying beds

Within the upper part of the Wig Bâch Formation is the Trwyn Cam Member, the higher of two arenaceous units mentioned above. This distinctive lithostratigraphical unit was traced for over a kilometre from Trwyn Cam along the west side of Aberdaron Bay. It is best exposed along the coast south of Trwyn Cam [1663 2600] and at the back of Porth Meudwy [1635 2551]. Exposures of concretionary sandstones on Ebolion [1889 2514], at the southern end of Trwyn y Penrhyn on the east side of Aberdaron Bay, are tentatively correlated with the Trwyn Cam Member. The base of the member grades up from finer, more argillaceous sequences with *Tetragraptus reclinatus*, recorded from dark grey siltstones exposed at this level in the valley leading to Porth Meudwy (Nant y Porth Meudwy of Beckly, 1985). Inland it can be traced for several hundred metres south-west from Porth Meudwy as a series of small exposures along a ridge, but it is lost (presumably by faulting) before the coast at Parwyd is reached.

The Trwyn Cam Member consists of about 35 m of bioturbated, grey, fine-grained sandstone. Individual sandstone laminae are separated by darker siltstone intercalations, the relative abundance of these two sediment types varying through the sequence. The sediments show scoured surfaces and masses of vertical burrows, many of which show meniscus-fill structure. Large calcareous concretions of diagenetic origin weather out to produce prominent oval hollows in the cliff face. Towards the top of the sequence there is a bed of structureless grey cherty siltstone interpreted as a tuff. The contact between the Trwyn Cam Member and the overlying beds is cut out by a low angle fault in Porth Meudwy, although the sandstones below this fault already include much argillaceous material. However, the contact is exposed further north along the coast in the small cove north of Ynys Piod [1653 2583].

Above the Trwyn Cam Member lie about 25 m of dark grey, laminated siltstone containing thin (up to 50 mm) sandstone beds. These sediments lack the intense bioturbation of the Trwyn Cam Member and show small-scale disruption produced by syn-sedimentary deformation, a feature not common elsewhere in the Wig Bâch Formation. Examples are exposed on the north-east side of Porth Meudwy [1638 2559]. This synsedimentary disruption may have been related to active faulting, which was responsible for the influx of extremely coarse sediments in the overlying formation.

Porth Meudwy Formation

The Porth Meudwy Formation is a most distinctive sequence of extremely coarse, matrix-supported conglomerates interbedded with numerous coarse-grained sandstones. It is 25 to 35 m thick and was traced only between the north side of Porth Meudwy [1642 2560] and the pre-

Plate 10 South-easterly dipping turbiditic sandstones of the Porth Meudwy Formation at Bau Ogof-eiral [1563 2404]. The crag in the middle left distance is Parwyd Gneiss. The cliffs in the background expose Llŷn Shear Zone metasediments thrust over Arenig rocks. View looking west. (A15016).

cipitous coast at Bau Ogof-eiral [1563 2404], a distance along strike of 1.75 km. Although conveniently exposed at low tide in Porth Meudwy, the exposures here are somewhat complicated by faulting. Better and more spectacular exposures exist beneath the dolerite sill along the cliffs west of Pen y Cil, at and immediately east of Bau Ogof-eiral (Plate 10). The coarsest beds contain metre-size boulders of Monian rocks equivalent in lithology to the phyllitic and finely schistose mylonites of the Llŷn Shear Zone, now exposed on the other side of Parwyd. The great size of these boulders, along with evidence such as intraformational rip-up clasts and synsedimentary slumping and folding, indicates rapid deposition under high energy submarine conditions. Many of the sandstone units show graded bases and typical Bouma sequences, indicating deposition by turbidity currents.

The sediments of the Porth Meudwy Formation may be grouped into four broad facies-types. Facies A is represented by coarse-grained turbidites, B by medium-grained turbidites and C by fine-grained turbidites with argillaceous intercalations. Facies D is represented by very coarse debris flows and slumps. The base of the Porth Meudwy Formation is placed at the base of the first coarse clastic bed (Facies A or D, see below). The top of this formation is taken at the top of the last coarse clastic bed.

Facies A: Coarse-grained turbidites

Three subfacies have been recognised:

A_1 Sheet-like, massively bedded turbidites which represent facies classes S_1 to S_3 of Lowe (1982) and classes A and B of Bouma (1962). Each bed is characterised by a scoured base, with or without flute marks, overlain by coarse-grained sand and gravel, above which are stratified (S_1) and graded-stratified (S_2) sediments, and finally dish and pipe structures (S_3) with a flat top. They are typically found as fining-upward cycles, although thin, coarse lenses are common within these units.

A_2 Channelled turbidites with limited lateral extent of up to 10 to 15 m, but with internal structure similar to subfacies A_1. Typically, they are lens-shaped in cross section with feather edges either side extending into interchannel deposits (subfacies A_3) and commonly overlain by finer-grained and thinner-bedded sandstone with planar and ripple cross-stratification, deposited either during or shortly after channel abandonment.

A_3 Thin sandstones and siltstones interpreted as thinly bedded crevasse splay and interchannel deposits. Although this subfacies is neither thick bedded nor always coarse grained, it is only developed in association with sheet and channelled subfacies (A_1 and A_2). These sediments are commonly affected by synsedimentary slumping (subfacies D_1) or postdepositional convolution, probably due to the high concentration of hemipelagic material.

The clasts within Facies A are very varied and include both intraformational sandstone and mudclasts, as well as extraformational pebbles of quartzite and mylonitic metasediment derived from the nearby Gwna Mélange and Llŷn Shear Zone. Subfacies A_1 and A_2 represent sheet-like and channelled beds respectively, deposited by turbidity currents with a high concentration of coarse and very coarse material. The sheet-like (A_1) beds are interpreted as deposits of turbidity flows that spread widely across the submarine fan. The smaller, more restricted channelled flows (A_2) are seen to cut into sheet-like beds, and were deposited in association with subfacies A_3. Primary current lineations show a NW–SE orientation that is consistent with a sediment source from the Monian basement to the north-west. A small number of planar, cross-stratification foresets show no strongly preferred orientation, although most dip broadly to the south-east, again suggesting dominant current flow from the north-west.

Facies B: Medium-grained turbidites

Two subfacies have been recognised:

B_1 This subfacies typically shows Bouma classes C to E as coarsening-upward cycles with parallel-bedded, fine-grained sandstones and a few hemipelagic and sometimes tuffitic siltstones and mudstones. The subfacies commonly coarsens up into thick-bedded, medium-grained sandstone with planar and trough cross-stratification.

B_2 This subfacies is similar to B_1 but without medium-grained, thick-bedded sandstone; it typically has a higher proportion of thin tuffitic and hemipelagic mudstone.

Subfacies B_2 represents the introduction of fine-grained clastic detritus into the dark mud and silt that dominated the background sedimentary environment characteristic of most of the Arenig succession in this district. The progradation of the clastic detritus progresses from B_2 to B_1, within which more thickly bedded sandstones are found. Facies B broadly records the action of relatively minor and more dilute turbidite flows across the area, both immediately prior to, and during the waning of the main phase of coarse clastic input. The orientation of primary current lineations suggests a dominantly NNE–SSW trend, which is the probable approximate local axial trend of the Arenig marine basin.

Facies C: Fine-grained turbidites

This facies is more thinly bedded than facies A and B, and is also recognised by lateral and vertical regularity of bedding and structures; it represents classes D and E of the Bouma (1962) cycle. Parallel laminated or bioturbated siltstone with hemipelagic intercalations dominate; it shows only very gradual changes in bed thickness and sandstone: mudstone ratio.

Facies D: Slump/debris flow deposits

This facies has two subfacies:

D_1 Slump horizons with pervasive disruption of beds into folds, thrusts, balls and hooks. The fold axes in these horizons are fairly reliable palaeoslope indicators and show preferred orientations towards N131° and N081°. These slumps are interbedded with Facies A and subfacies D_2.

D_2 Debris flows occurring as beds of structureless, disorganised, matrix-supported sedimentary slurries which support clasts of various lithologies (Plate 11). These clasts include reworked intraformational mudstone, siltstone and sandstone (and previously reworked debris flow), and extraformational Monian quartzite and mylonitised metasediment. Bed thickness varies from 0.5 to 2.5 m, and clast size varies greatly, reaching up to 3 m in maximum diameter; these huge clasts were locally derived from the nearby Llŷn Shear Zone, now exposed on Trwyn Bychestyn. Above a negatively graded scoured base, the flows show generally normal grading upwards into mud, although some have a sharp top and others grade up into cross-bedded turbidites (subfacies A_2) due to further dilution of the slurry. D_2 debris flows are more abundant throughout the Porth Meudwy Formation than subfacies D_1.

The Porth Meudwy Formation has not been recognised anywhere east of Aberdaron Bay. This may be due to lack of exposure, a proposition supported by the fact that the highest beds exposed on the mainland on the east side of Aberdaron Bay, at Ebolion, bear a close resemblance to the Trwyn Cam Member of the Wig Bâch Formation. If this correlation is correct, then the Porth Meudwy Formation may crop out somewhere on the sea bed between the mainland and the island of Ynys Gwylan-fawr. Alternatively, the turbiditic sediments of the Porth Meudwy Formation may have been narrowly confined, especially if the Daron Fault possessed a synsedimentary topographical expression.

Aberdaron Bay Group (undivided)

West side of Aberdaron Bay

The sequence lying immediately above the Porth Meudwy Formation is well exposed on the west side of Aberdaron Bay, around Porth Meudwy [1646 2561; 1638 2540] and at Bau Ogof-eiral [1567 2406]. It comprises thin pale grey to white tuffs and ashy and cherty, tuffaceous nodular sandstones and siltstones, as well as bioturbated grey sandstones and siltstones more typical of the Wig Bâch Formation below. At one level in the exposures

Plate 11 Overturned slump hook within debris flow (facies D_2) in Porth Meudwy Formation at Bau Ogof-eiral [1563 2404].

at Porth Meudwy there is a nodular, silty, pisolitic ironstone. Cherty sandstones, associated with dark micaceous siltstones, also crop out in the Afon Daron gorge, adjacent to the Daron Fault. The sediments above the Porth Meudwy Formation have yielded a somewhat indefinite graptolite fauna that Beckly (1988, p.325, location 12) suggested was of Fennian age. New work favours an early Llanvirn age (see p.32).

The top of the Porth Meudwy Formation is placed at the top of the highest coarse (Facies A) bed. At the type locality, on either side of Porth Meudwy, this bed is an approximately 0.1m-thick, matrix-supported, conglomeratic sandstone with rare oncoidal and phosphatic debris. This is overlain by 2 to 3 m of bioturbated, commonly nodular, fine-grained sandstones and siltstones, marking a relatively brief return to sedimentary conditions similar to those recorded by the more arenaceous parts of the Wig Bâch Formation. These sediments are overlain by a distinctive pale grey, banded tuff up to 0.56 m thick. Above the tuff, the sediments have a distinctive volcaniclastic component that is the hallmark of this part of the group. Bioturbated, ashy, fine-grained sandstones, interbedded with pale, thin ash bands, characterise the lowest 2.8 m of the succession.

An overlying dark grey, silty, nodular, pisolitic ironstone up to 150 mm thick forms a distinctive marker bed traceable across the bay (Plate 12). It forms a well-exposed nodular bedding surface on the north side of Porth Meudwy, and also crops out in the cliff adjacent to the dolerite sill south of Porth Meudwy. The iron-rich ooids are seen to lie in thin millimetre-scale laminae interbedded with organic-rich mudstones and siltstones similar to the underlying bed. The organic material includes graptolites and abundant, mostly fragmented chitinozoans. The ironstone also includes laminae of 100–150μm sand grains. The proportion of ooids increases upwards to 80 per cent, by increasing thickness and frequency of the ooidal laminae. The ooids are variable in size, although those in individual laminae tend to be well sorted. They are typically about 400μm in diameter, but can reach 1000μm. Larger, coated intraclasts (oncoids?) up to 6 mm in diameter occur in some laminae. The ooids are phosphatised in places, and in some of the grainstone laminae there is also a phosphatic cement. The lamination is disturbed in places by bioturbation, with burrows of indeterminate type. The ironstone bed is sharply overlain by a volcaniclastic sandstone which, near its base, includes ferruginous ooids and intraclasts of siltstone. Although the bulk composition of the ironstone as a mixture of mudstone/siltstone and ferruginous allochems might appear to class this deposit as an ooidal mud-ironstone, it is more a fine interlayering of clastic siltstones and ooidal grain-ironstones. The differentiation of the rock into thin ooidal grainstone laminae and organic siltstones suggests that the ooids have been derived, although they do not appear to have been transported far.

Above the ironstone is a distinctive series of ashy, feldspathic sandstones with calcareous concretions. These sandstones are best exposed on the north side of Porth Meudwy [1646 2560] where they are 3 to 4 m thick and overlain by a sequence of over 12 m of cherty, commonly nodular sandstones, siltstones and cherts. A volcaniclastic component is present throughout this sequence (Plate 13). On the south side of Porth Meudwy, only 4.8 m of sediments lie between the ironstone and an overlying sill. The difference in exposed thicknesses between north and south sides is presumably due to some degree of intrusive cross-cutting exhibited by the 'sill', with the dolerite lying some 7.2 m higher in the sequence on the north side of the bay. The fossil assemblages collected by Beckly (1988) and Young (1991, p.325) tentatively suggest a Llanvirn age for at least part of the sequence above the Porth Meudwy Formation.

There are several differences in the Ordovician sequence exposed above the Porth Meudwy Formation on

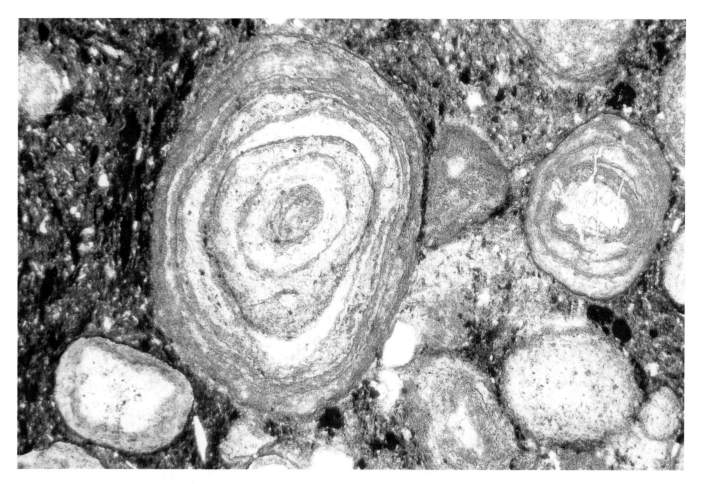

Plate 12 Phosphatic ooids within muddy ironstone exposed on the north side of Porth Meudwy [1645 2560]. The high sphericity reflects the pre-compactional diagenetic replacement of iron silicate by phosphate minerals. Field of view = 1.5 mm side to side. Plane polarised light.

the coast at Bau Ogof-eiral, only 1.5 km to the south-west. Here, the top of the Porth Meudwy Formation is represented by a highly distinctive 0.43 m-thick, matrix-supported, conglomeratic, oncoidal ironstone. This bed is interpreted as a resedimented mass-flow deposit that includes reworked ferruginous and phosphatic material. These elements are mixed with large quantities of coarse-grained sandstones derived from the local Monian se-quences. In places, the coarser-grained parts of the beds are cemented by ferruginous carbonate (Fe-dolomite), reducing the initially high porosity. The resedimented oncoidal ironstone is overlain by dark grey, ferruginous siltstone (0.09 m) containing clasts winnowed from the underlying bed. In the adjacent muddy interbeds, there is commonly a rhombic carbonate cement phase, and one of these mudstones shows a fine ferruginous lamination that may be of stromatolitic origin.

As at Porth Meudwy, sedimentation then briefly resumed a character more typical of the Wig Bach Formation, although here only 0.6 m of bioturbated fine-grained sandstone and siltstone occur beneath the tuff (as opposed to 2.3 m south of Porth Meudwy and 2.8 m north of the bay). The pale banded tuff forms a distinctive 0.4 m-thick marker horizon running through the cliffline. Above it is a bioturbated fine sandstone, 0.5m thick, followed by another 0.4 m-thick pale tuff band, overlain by 0.23 m of ashy sandstones. Above this, however, the strong volcanic influence is virtually lost in a 9.7 m sequence of bioturbated grey sandstones and siltstones. However, the highest 2 m exposed beneath the Pen y Cil dolerite sill show the return of a volcaniclastic component, for the sandstones are often cherty and ashy, and contain phosphatic nodules, particularly in the upper part of individual beds. The differences in detailed lithostratigraphy between Porth Meudwy and Bau Ogof-eiral suggest that rapid local variation existed across the fault-influenced Ordovician sea floor in the Aberdaron area.

Afon Daron Gorge

The Afon Daron gorge around Pandy-Bodwrdda [1895 2732 to 1919 2771] provides the best Ordovician exposures inland from Aberdaron. Most of the rocks dip at about 60° to the south-west and are typically strongly cleaved, dark grey siltstones and mudstones, although

Plate 13 Volcaniclastic sandstone from immediately above an ironstone bed [1645 2560] (see Plate 12) on the north side of Porth Meudwy. The sandstone is particularly rich in fresh, euhedral plagioclase crystals (centre). A – plain polarised light; B – crossed nicols. Field of view = 1.5 mm side to side.

siltstones with fine-grained sandstones crop out low in the sequence on the west side of the stream [1918 2768]. Within this generally lithologically uniform succession, a sequence of hard, cherty siltstones (the 'Daron Cherts' of Matley, 1932) crops out; these beds are considered to be air-fall tuffitic rocks. They are exposed in the bed of the Afon Daron [1907 2745]. Downstream [1897 2740], above these cherty rocks, *Platycalymene* sp. and pendent and biserial graptolites have been collected (Beckly, 1988 p.328). This fossil assemblage is of Llanvirn age and suggests that much of the unexposed ground east of Aberdaron, above the dolerite sill and west of the Daron Fault, is underlain by Llanvirn strata.

Ynys Gwylan-fawr

Sediments of the Aberdaron Bay Group lie between two basaltic units on the island of Ynys Gwylan-fawr, 600 m south-south-west of the headland of Trwyn y Penrhyn [1852 2465]. They comprise cherts overlain by dark grey, tuffaceous sandstones and mudstones that form a screen up to 8 m thick between pillowed basalt below and peperitic dolerite above. This sediment screen can be traced for 360 m from the north-east [1852 2466] to the south-western tip of the island [1829 2442]. These sediments appear to lie above the highest strata seen on Trwyn y Penrhyn, and the presence of abundant volcanic detritus is suggestive of a Fennian or (more probably) Llanvirn age

Nant y Gadwen

The narrow valley of Nant y Gadwen strikes south-west, reaching the coast south of Llanfaelrhys [2100 2653]. It provides one of the few extensive inland exposures of Ordovician rocks, and includes one of the two main manganese mines in the district (chapter six). Most of the western side of the valley consists of mudstones and siltstones dipping to the south-east, overlain by a horizon of pale weathered dolerite and baked siltstone (exposed at [2104 2658]). Beckly (1985, 1988) recorded a Fennian fauna that includes several graptolites, such as *Didymograptus distinctus, D. hirundo, D. uniformis lepidus, Isograptus caduceus gibberulus* and *Pseudisograptus angel*, and the trilobites *Pricyclopyge binodosa eurycephala* and *Microparia broeggeri*. On the opposite side of the same valley, grey siltstones also dip moderately south-eastwards. Here, Beckly (1985, 1988) recorded the trilobite *Cyclopyge grandis grandis* [2116 2668] and interpreted these sediments as Whitlandian, that is, older than those on the west side. In South Wales the range of *C. grandis grandis* extends from the upper Whitlandian to the lower Fennian (Fortey and Owens 1987), and in the middle Fennian it is succeeded by the subspecies *C. grandis brevirhachis*. Since all the taxa recorded from the west side of the valley that are known also in South Wales are middle Fennian, the faunal evidence suggests that there is no simple succession from west to east.

There is a marked difference between the exposures described above and those a little to the north-east. Here, across a north-west-trending fault, the rocks display a low to moderate (30° to 50°) south to south-westerly dip. A poorly exposed dolerite sill is overlain by grey mudstones containing the trilobite *Merlinia selwynii* [2122 2685], indicating a Moridunian age (Beckly, 1988). Across a gap in exposure, with no change in dip, tuffaceous mudstones, siltstones and a prominent bedded tuff crop out at [2122 2684], from which Beckly (1985) recorded a varied fauna of trilobites and graptolites, including *Didymograptus distinctus, D.* cf. *extensus linearis* and *Pseudisograptus* cf. *dumosus*. These are interpreted as of Fennian age (Beckly, 1985; but not referred to in Beckly, 1988). Continuing south west, apparently overlying these Fennian strata, are manganese-stained laminated siltstones which have yielded a Whitlandian fauna that includes *Bohemopyge scutatrix* and *Cyclopyge grandis grandis* (Beckly, 1985). Thus, across a distance of some 100 m rocks interpreted palaeontologically as representing all three Arenig stages are exposed, with the youngest faunas (Fennian) occurring above the oldest (Moridunian) but beneath those of intermediate age (Whitlandian). Because of the obvious complexity of this area, combined with the lack of exposure away from the valley sides, it has not proved realistic to determine a detailed mappable lithostratigraphy at the 1:10 000 scale and the area is referred to on the accompanying map as Aberdaron Bay Group (undivided).

Borehole data between Nant y Gadwen and Benallt

Ordovician rocks in this area were encountered in boreholes drilled east of Nant Mine, recorded by Groves (1952, pp.315–318) and Beckly (1985, p.85). Additional boreholes have more recently been drilled at Tyddyn Meirion [2222 2753], of which Rhiw Boreholes 1A and 1B (Brown and Evans, 1989) are of particular interest in containing ironstone. The ironstone lies between two dolerite sills and is about 3 m thick. It is very well sorted and contains ooids, 600 to 1000μm in diameter, which are of high sphericity and mostly composed of magnetite, whereas the cement is chloritic, with some phosphate. Laminae containing slightly differing sizes of ooids are visible in thin section, and this probably represents a primary bedding fabric. This ironstone is a reworked deposit, which may reflect input of ooids from a source similar to that which contributed to the sediments of the Porth Meudwy Formation.

Below this ooidal ironstone in borehole 1B, there is a very poorly sorted sediment, which contains rare ooids (250μm diameter) and abundant oncoids (up to 6 mm diameter), as well as clastic coarse-sand-grade material (600 to 800μm diameter), including mylonite fragments, fresh euhedral volcanic quartz grains and chloritised sand-grade grains of uncertain origin. Some of the probable oncoids are overgrowths on chloritised lithic grains, which appear to be altered basalts. The material from this facies occurs in the 0.6 m between the base of the ironstone and the top of the underlying sill. The diagenesis of the sediment included the generation of large carbonate rhombs, both in the matrix and within the oncoids. The rhombs, now pseudomorphed by chlorite, are similar in size to those seen in the ferruginous debris flow at the top of the Porth Meudwy Formation at Pen-y-Cil. A tentative correlation is made between the ironstones proved at Tyddyn Meirion and the lower Llanvirn (or possibly latest Arenig) ironstones on the west side of Aberdaron Bay.

Benallt

Ordovician rocks are still exposed in the abandoned Benallt mine workings on the south-west side of Mynydd Rhiw [2219 2828]. The exposures lie within and on either side of a north–south cleft in the hillside. The west face of the workings exposes heavy, black, massive and locally pisolitic manganiferous ore. It is associated with manganiferous and pyritic mudstones and siltstones with a Fennian fauna, including *Cyclopyge grandis brevirhachis* and *Pseudotrigonograptus minor* (Beckly, 1988), red cherts and pale, ashy volcaniclastics. All the exposures are cut by numerous faults. At the northern end of the excavations, grey siltstones and mudstones dip to the south-east at 16° to 48° and form upright folds that plunge south at up to 46°. A steep, north–south-striking cleavage is common throughout the exposures, especially in the northern end of the mine.

The east face of the cleft exposes grey, manganiferous, Moridunian siltstones and mudstones, with pronounced pale weathering produced by the migratory leaching of manganese and iron from the sediment. Most joint surfaces are coated with the remobilised manganese and all exposures are pervasively affected by tectonic disruption, expressed primarily as an anastomosing network of shiny, slickensided surfaces. Most of the sediments exposed in the workings are interpreted as slivers of Moridunian sediments caught up within the Benallt fault zone. Towards the southern end of these eastern exposures, the sediments contain a sill of highly vesicular, deeply weathered dolerite [2221 2812]. At the far southern end of the workings are piles of fresh pisolitic manganese ore left behind after the last phase of mining. These are overlain by scree slopes full of volcaniclastic chert, which is exposed in situ close to Cadwgan [2227 2798]. Between this chert and the main workings are vesicular, pillowed dolerites (the Clip Lava, see p.44), which are exposed in the hillside at [2229 2810] and can be traced north-north-eastwards to Clip y Gylfinhir [2240 2847]. The Arenig succession on both sides of the fault zone contains numerous dolerite sills. The most prominent of the few exposures of these dolerites on the west side of Mynydd Rhiw occurs 320 m north-west of the northern end of the Benallt workings, at Baron Hill [2191 2847].

Mynydd Penarfynydd

At Mynydd Penarfynydd in the south-eastern corner of the district, dark grey micaceous mudstones and siltstones with pale tuffaceous beds are exposed beneath the basaltic chilled margin of the Mynydd Penarfynydd layered intrusion (Matley, 1932) [2148 2590]. The Ordovician sequence is cut by two northward-dipping reverse faults beyond which the base of the overlying sill reaches sea level, eliminating exposure of the sedimentary rocks.

Pale volcaniclastic siltstones are particularly prominent in the upper 3m of exposure behind Penarfynydd Farm [2188 2663]. The volcaniclastic sediments at this locality overlie up to 4 m of fossiliferous grey siltstones, which yield a Llanvirn fauna of graptolites and trilobites. Other exposures are very scarce, but may be found at the base of the sill around the summit of Mynydd Penarfynydd [2200 2659], in an old quarry at [2157 2683] and between two dolerite sills on the coast at [2116 2627].

Baked grey mudstones, siltstones and sandstones, dipping south-east at 30° to 40°, are exposed on the cliffs [2230 2625] east of Mynydd Penarfynydd, above the layered intrusion. These sediments are commonly nodular, pyritic, manganiferous and bioturbated, and they locally show wave ripples. They appear to lie above the Llanvirn strata exposed beneath the Mynydd Penarfynydd layered intrusion and are therefore deduced to be the youngest strata present in the district. They show strong hornfelsing close to the top of the Mynydd Penarfynydd layered intrusion, where they are bleached, hardened and locally pyritised.

FOUR
Intrusive igneous rocks

The post-Precambrian intrusive igneous geology of the district is dominated by a suite of south-easterly dipping dolerite intrusions emplaced into the Arenig and Llanvirn sedimentary rocks, and the Mynydd Penarfynydd layered intrusion that lies concordantly within Llanvirn strata. The dolerites mostly take the form of transgressive sills, although in places they show a dyke-like form, cutting the bedding at steep angles. These sills appear to climb the stratigraphical sequence from east to west, the lowest intruding Moridunian strata in the east (e.g. at Porth Cadlan) and Llanvirn strata in the west (e.g. at Porth Meudwy). On the island of Ynys Gwylan-fawr, dolerites are associated with pillow lavas and peperites. Pillowed textures are also found in a lava occurring east of the manganese mines; this was named the 'Clip Lava' by Matley (1932).

Narrow feldspathic dykes cut the Arenig strata at one locality on the coast east of the Daron Fault [1856 2552]. In addition, over 35 altered dolerite dykes of probable Palaeozoic age intrude the Gwna Mélange along the coastline north-east of Porth Oer. Similar basic dykes exposed inland cut exposures of both the Llŷn Shear Zone and the Sarn Complex. Several olivine dolerite dykes of Tertiary age intrude both the Gwna Mélange and the Llŷn Shear Zone; they crop out at three localities on the coast of Bardsey, along the south-west coast of the mainland at Trwyn Bychestyn and Trwyn Maen Melyn, and at four localities on the north-west coast.

DOLERITE

Only one dolerite sill (the Pen y Cil Sill) is exposed west of the Daron Fault, whereas many crop out within the Ordovician succession to the east of this fault. The Pen y Cil Sill is a thick, south-easterly dipping intrusion named after its type locality on the south-west coast [158 240]. The top of the sill is nowhere exposed, but the excellent exposures at Pen y Cil show that it is at least 150 m thick. The base is exposed at several localities on the coast [1568 2407; 1635 2517; 1640 2539; 1648 2560], where it is everywhere concordant with the underlying Ordovician strata. Immediately beneath the basaltic chilled lower margin of the sill, the sediments are baked and bleached to produce a narrow but distinctive, white-weathering, porcellanous zone. Inland, the sill has been traced westwards from the old quarry north of Aberdaron [1747 2690] towards the Daron Fault, against which the dolerite outcrop is abruptly truncated. Weathered dolerite is exposed in the track leading south towards Pandy-Bodwrdda [1893 2745], in the banks of the Afon Daron [1900 2745] and in the ground immediately to the south [1902 2742].

The lowest, and thickest dolerite intrusion east of the Daron Fault trends parallel to the coast north-eastwards from Trwyn Bychestyn [1914 2545] to Porth Ysgo (Plate 14). Matley (1932) named this intrusion the Gallt y Mor Sill. It is about 140 m thick and was probably emplaced at the same time as the Pen y Cil Sill, which is the lowest dolerite sill in the western part of the district. It is likely that these two sills were linked, with the Gallt y Mor intrusion climbing westwards to higher stratigraphical levels to feed the Pen y Cil Sill. The base of the Gallt y Mor intrusion is discordant where it is exposed on the coast, just above sea level immediately east of Ogof Lleuddad [1908 2552]. The dolerite steps up across the bedding in the underlying laminated siltstones and is clearly transgressive in nature. An isolated continuation of this intrusion occurs on Carreg Gybi [1904 2527]. The next intrusion higher in the sequence from the Gallt y Mor Sill is the Maen Gwenonwy Sill, which is well exposed at its type locality [2010 2596]. The dolerite on Carreg Chwislen [1998 2580] is correlated with this sill.

A detailed magnetometer survey was carried out in 1984 over the ground north-eastwards from Porth Ysgo to Mynydd Rhiw (Brown and Evans, 1989). The Gallt y Mor Sill is significantly magnetised and could thus be readily traced across this drift-covered area. The results of the survey show that the Gallt y Mor Sill is next exposed at Baron Hill [2191 2846], where it was named the Footwall Sill by Groves (1947, 1952) because of its relationship to the workings of the Benallt manganese mine. Brown and Evans (1989) also suggest that the Maen Gwenonwy and Gallt y Mor sills might merge inland from Porth Ysgo, the former being thus a westward-climbing branch of the latter.

A characteristic feature of these dolerite intrusions is a well-developed chilled margin, commonly with oval, now calcite-filled vesicles produced by steam escaping from the baked sediments below. Good examples of this distinctive texture are seen at the base of both the Pen y Cil and Gallt y Mor intrusions; one of the best is exposed on the north side of Maen Gwenonwy [2010 2597] (Plate 9), where the base of the sill shows a chilled basaltic contact conformable with finely spotted, baked, grey mudstone. The chilled margin displays a spherulitic devitrification texture that can be traced 2mm in from the contact. There are also mud injection features and a vesicular texture, with each vesicle oriented at 90° to the margin. On Pen y Cil itself, the base of the sill displays pipe-like lines of vesicles rising into the igneous rock from the sediment beneath [1566 2410]. It is concluded that these sills intruded wet sediments, probably at shallow crustal levels. The transgressive contact at Ogof Lleuddad [1907 2552] supports the idea that the Gallt y Mor intrusion cuts up-sequence in a roughly westerly direction towards the higher intrusion levels of the Pen y Cil Sill seen west of the Daron Fault.

The dolerites generally preserve excellent ophitic and subophitic textures, with altered plagioclase crystals en-

Plate 14 Porth Ysgo [2069 2648] showing the highly vesicular top of the Gallt y Mor dolerite intrusion (lower left) overlain concordantly by sandstones which are strongly baked close to the sill. The sandstones dip to the south-east and are cut by minor faults. The thick, locally derived drift seen at the top right infills the valley eroded along the Porth Ysgo Fault. (A15027).

veloped by augitic clinopyroxenes. Ilmenomagnetite is usually partly altered to leucoxene, and all specimens examined were rich in secondary chlorite; sericite, albite, pumpellyite and epidotic minerals also occur. The weathered samples from inland exposures (e.g. near Pandy-Bodwrdda [1892 2746]) show more alteration and are stained brown by limonitic secondary products. In the old quarry at Baron Hill [2191 2846] on the west flank of Mynydd Rhiw, the thickness of the Gallt y Mor Sill is reflected by the coarse texture, with individual crystals of plagioclase and anhedral masses of augite reaching over 1.5 mm in size.

YNYS GWYLAN

During the course of this survey, new and interesting discoveries were made on the small island of Ynys Gwylanfawr [1840 2455]. Most of the rocks on this island are of igneous origin, with the exception of a south-easterly dipping screen of volcaniclastic sandstones and mudstones. These sediments separate two basic igneous units: a lower pillow basalt (>15 m thick) and an upper unit (>35 m thick) that includes pillow basalts, dolerites and mixtures of igneous and sedimentary material (peperites).

The lower pillowed basalt is continuously exposed along the north-west coast of the island, with the best examples of pillows in the south-west [1829 2445]. The pillows are right way up and young towards the south-east with an average dip of about 44°. Equally well-developed pillows, dipping in the same direction, are present in the upper unit, on the opposite coast of the island [1838 2446]. The largest pillows occur here and commonly reach a metre or more in diameter; they are underlain by a massive basaltic unit which shows a chilled margin with the underlying sediments. The upper unit of pillow lavas passes north-eastwards into a peperitic zone in which baked mudstones are intermixed with fragments of the basaltic igneous rocks. Peperitic textures are particularly well exposed in places along the south-east-facing cliffs [1847 2452]. Similar rocks occur along much of the north-eastern coast.

The Ynys Gwylan igneous rocks occur along strike from the Gallt y Mor and Maen Gwenonwy dolerite intrusion on the mainland. The presence of abundant pillows and peperitic textures indicates either extrusive and/or very shallow level intrusive conditions on Ynys Gwylan. The abundance of volcanic material in the Ynys Gwylan sediment screen, together with the consistent south-easterly dip on both mainland and island, suggests that these sediments are younger than the Arenig rocks exposed on the nearest headland on the mainland to the north (Trwyn y Penrhyn), and are probably of Llanvirn age. The age of the dolerites and basalts is not known, although the shallow intrusive or possibly extrusive textures suggest that there is not a large age difference between the sediments and igneous rocks. It is likely, therefore, that the dolerite sills in this district, as well as the Ynys Gwylan igneous rocks, are all of Llanvirn age.

THE CLIP LAVA

The Clip Lava is an igneous unit referred to by Matley (1932), who named it after a conspicuous exposure on Mynydd Rhiw at Clip y Gylfinhir [2239 2847]. The exposure was incorrectly described as a volcanic neck by Blake (1888). The unit on Mynydd Rhiw in part comprises pillowed basalts, with interstices infilled by cherty sediment, whereas in other areas the rock is clearly doleritic. Vesicles infilled with calcite or chlorite are abundant, and secondary pumpellyite occurs. The margins of this body are not exposed and it is therefore uncertain whether it is truly an extrusive lava or a shallow level intrusion. The unit is poorly exposed but can be best seen on the western side of Mynydd Rhiw at [2228 2809; 2230 2829].

The Clip Lava is also exposed near Llanfaelrhys [2168 2698], 1.3 km south-south-west from the exposures on Mynydd Rhiw. That this exposure is the Clip Lava has been demonstrated by drilling (Groves, 1952), which confirmed its position above the manganese mudstones, and by magnetometer survey (Brown and Evans, 1989), which confirmed its position with respect to the doleritic sill to the west. Here, south-eastward-younging basaltic pillow lava and hyaloclastites contain interstitial chert, and again show anchizonal alteration to a pumpellyite-bearing assemblage. Groves has suggested that this basalt may be equivalent to the basic igneous rock exposed on the south-east side of Nant y Gadwen [2109 2657]. Certainly they strike towards each other, although the latter exposures show an intrusive, doleritic rock that is quite different from the pillowed lavas on Mynydd Rhiw, to the north-east. However, the fact that the Clip Lava is extremely variable in texture and possesses features compatible with a shallow-level intrusive body is considered enough justification to correlate it with the Nant y Gadwen intrusion to the south-west. Present exposure is not good enough to prove whether all these basaltic rocks are shallow intrusions, extrusions, or some combination of both.

Surface exposures on the coast at Porth Ysgo [2113 2626] show that at least two dolerite sills lie between the Clip Lava and the layered sill of Mynydd Penarfynydd (Matley, 1932). It is likely that several more basic sheets lie within the Llanvirn strata beneath the Mynydd Penarfynydd layered intrusion, although there is neither exposure nor subsurface data to prove this.

THE MYNYDD PENARFYNYDD LAYERED INTRUSION

The Mynydd Penarfynydd intrusion, which crops out in the south-eastern corner of the district, is a major part of the Rhiw Igneous Complex and intrudes Llanvirn sedimentary rocks. Originally described as a 'greenstone' (Sharpe, 1846; Ramsey, 1866; Tawney, 1880; Bonney, 1881 and Teall 1888), it has been the subject of detailed petrographic studies by Harker (1888, 1889), who described the rocks as hornblende-picrites and hornblende-diabases. Hawkins (1965) demonstrated the igneous cumulate nature of the layering, and Cattermole (1976) interpreted the intrusion as being derived from the fractional crystallisation of a hydrated alkali-olivine basalt magma. A suite of whole rock geochemical analyses from this intrusion is given in Table 3 (additional analyses are provided by Cattermole, 1976). The marked geochemical variations have been interpreted as primarily a result of igneous differentiation acting upon a number of separate intrusions of magma (Cattermole, 1969).

The intrusion forms the hill of Mynydd Penarfynydd, which rises precipitously from sea level to a height of 177 m. The rocks dip at about 40° to the south-east, so that the full sequence may be traversed eastwards from Trwyn Talfarach [2148 2582]. The baked sediments beneath the sill form a distinct ridge at the north end of Mynydd Penarfynydd [2220 2665] and form the highest point on the hill [2201 2658]. The total thickness of the intrusion is about 150 m, with most of the lowest 100 m comprising hornblende-picrite. The dark, dense picrite is extremely well exposed on Trwyn Talfarach and weathers to produce a highly distinctive honeycomb texture (Plate 15). At the base of the picrite is a complex chilled zone [2147 2582] about 10 m thick that grades upwards from a glassy, basaltic margin (1.5 m), through hornblende dolerites into fine-grained hornblende gabbro (with an increased quantity of olivine and phlogopite). Two prominent leucocratic bands occur in this lower margin, although the rocks are generally not well layered. Secondary alteration along joint surfaces in this lowest part of the intrusion has produced chlorite and actinolite.

The picrites (E^P) vary from massive to strongly layered. They are dominated by cumulus olivine crystals plus minor chrome-spinel, intercumulus brown pargasitic hornblende, phlogopitic mica and rare plagioclase feldspar (bytownite to labradorite). Resting upon the picrites, with a well-defined, sharp contact, are banded leucogabbros (E^L); these rocks are spectacularly exposed in a south-west-facing cliff [2168 2573]. The leuco-

Table 3 XRF whole rock analyses for lithologies within the Mynydd Penarfynydd layered intrusion.

	Lower chilled margin	Picrite		Leucogabbro	Melagabbro	
Sample No.	M1	M2	M3	M4	M5	M6
Height from base (m)	0.2	20	80	105	125	130
SiO_2	46.7	39.6	38.7	45.7	38.8	40.5
TiO_2	1.19	0.52	0.48	0.53	2.42	2.26
Al_2O_3	14.9	5.79	5.60	18.59	12.9	14.16
Fe_2O_3	11.3	12.5	15.7	6.20	18.3	16.6
MnO	0.17	0.18	0.23	0.10	0.20	0.18
MgO	9.15	31.1	27.3	9.73	7.99	6.90
CaO	8.90	4.28	5.45	13.7	14.4	14.3
Na_2O	2.69	0.43	0.34	1.44	0.75	0.99
K_2O	0.16	0.18	0.16	0.36	0.47	0.47
P_2O_5	0.13	0.05	0.06	0.07	0.17	0.11
Total	95.4	94.6	93.9	96.4	96.3	96.5
Ba	82	54	43	61	131	142
Zn	88	62	70	39	52	57
Ni	111	764	485	102	20	11
Cr	402	2309	1581	266	61	43
Ce	55	17	33	17	23	81
V	242	118	132	213	848	745
Rb	11	14	10	29	15	18
Sr	197	84	70	179	73	165
Y	22	12	10	18	18	18
Zr	74	36	26	72	37	38
Ap	0.30	0.12	0.14	0.17	0.39	0.25
Il	2.24	0.98	0.90	1.01	4.53	4.24
Mt	4.09	4.49	5.60	2.24	6.53	5.92
Or	0.93	1.07	0.93	2.11		2.73
Ab	22.52	3.57	2.83	12.11		2.43
An	27.85	13.17	13.07	42.93	29.80	32.30
Di	12.23	5.91	10.54	19.35	30.65	22.82
Hy	12.89	4.71	3.51	1.55		
Ol	12.37	60.60	56.60	14.93	18.16	14.77
Ne					3.38	3.15
Lc					2.14	
Cs					0.63	
Total	95.41	94.62	93.92	96.40	96.30	96.46
A	11.57	1.34	1.11	9.77	4.13	5.46

Major elements in oxide wt%.
Trace elements in ppm.
FeO/Fe(t) estimated as 0.75 for CIPW Norm calculation.

gabbros are about 9m thick and are dominated by plagioclase, with augitic clinopyroxene and minor pargasitic hornblende and olivine. This feldspathic unit is overlain by about 13 m of hornblende olivine gabbro (E), commonly with pegmatitic textures near the base. The lower contact of this gabbro appears to erode the underlying leucogabbro.

A prominent, dark band of melagabbro (E^M) (Plate 16), about 10 m thick, lies above the hornblende olivine gabbro; it is easy to trace down the cliff to sea level [2170 2585 to 2180 2576]. It is rich in both augitic clinopyroxene and magnetite, with plagioclase, minor olivine and pargasitic brown hornblende. Above the melagabbro, the rocks become progressively more differentiated through feldspathic, ilmenitic gabbroic (E) and dioritic (H^L) rocks, to a distinctive granophyric lithology at the highest stratigraphical level of exposure. All these 'upper zone' rocks are affected by secondary alteration, and commonly display blotchy, mottled textures. An attempt has been made on the map to differentiate between a lower hornblende gabbro and an upper meladiorite, the latter being characterised by blotchy texture. It is likely, however, that these two rock types may represent one gradational unit that has been more pervasively altered in its upper part (they are equivalent to the Lower Hornblende Gabbros of Cattermole, 1976; and the Metadiorite of Hawkins, 1970). The mottling is a result of secondary alteration to chlorite and zeolites; other secondary minerals present include calcite and albite.

The mottled rocks mentioned above grade upwards into very pale, cream-coloured rocks rich in albite and quartz, with minor hornblende, epidote, sphene and ilmenite. These granophyres are well exposed at the base

Plate 15 Picrite within the Mynydd Penarfynydd layered intrusion at Trwyn Talfarach. Weathering of the cumulate texture has produced a distinctive honeycomb pattern. [2152 2580].

of the cliff and display distinctive sprays of chlorite that are particularly well exposed at [2190 2580] and [2200 2588]. Because the contact between these granophyres and the underlying metadiorites is gradational, these two rock units are not separated on the map. The baked sediments forming the roof of the intrusion are exposed in the small cove on the north-east side of Mynydd Penarfynydd [2226 2620], where blotchy, xenolithic metadiorite and granophyre are in contact with indurated Llanvirn rocks. Irregular felsitic and commonly pyrite-rich dykes and sills intrude the hornfelsed country rocks above the upper contact of the layered intrusion. Inland exposure of the layered intrusion is poor, with the exception of the leucogabbro which crops out as a line of crags [2165 2590 to 2210 2642].

OLIVINE DOLERITE DYKES (TERTIARY)

The mostly NW–SE-striking olivine dolerite dykes of Tertiary age are found intruding the Gwna Mélange (e.g. at Ogof Hir [1152 2226] on Bardsey Island) (Plate 17) and the Llŷn Shear Zone (e.g. those on Trwyn Bychestyn [1520 2427]). Other examples cut the Gwna Mélange on the coast both south [1380 2524] and north [1424 2631] of Mynydd Mawr, and on the coast west of Mynydd Anelog [1474 2733; 1485 2795]. The most northerly example exposed in the district occurs at the coast west of Porth Colmon, where a 0.9 m wide dolerite has been eroded to form a prominent gully running north-north-east out to sea [1936 3431]. This latter dolerite is one of only two examples where the structure of the country rock appears to have exerted a strong influence on the intrusion path, diverting it away from the normal NW–SE strike. The other is the 9m-wide, ESE-trending Ty Canol Dyke (Groves, 1947, 1952) which was proved in the workings of the Benallt Mine.

These dolerites mostly form only very thin intrusions, usually less than 1m in thickness (e.g. 0.9 m at Porth Colmon; 0.3 m north of Mynydd Mawr); they are commonly highly vesicular. The Porth Colmon dyke, for example, exhibits a weathered basaltic chilled margin, inside which is a prominent vertical band rich in calcite vesicles. Flow texture is sometimes exhibited, and is best seen at the margin of the dyke west of Mynydd Anelog [1474 2733], where it is defined by fine crystals of aligned plagioclase feldspar.

Plate 16 Banded melagabbro in Mynydd Penarfynydd layered intrusion. [2173 2580].

The dolerites differ from the other basic intrusions seen in the district, both in their field appearance and mineralogy. They usually form preferentially weathered gullies in the cliffline, and are brown weathered in hand specimen, although this belies a remarkable mineralogical freshness in thin section. This freshness is particularly well displayed by the plagioclase feldspars, although the clinopyroxenes are also unaltered. Significant alteration is restricted to the olivine crystals, the oxidation of which has released iron that is primarily responsible for the brown, limonitic staining of these rocks at outcrop. By comparison with similar intrusions elsewhere in Wales, it is likely that these dykes were intruded during early Tertiary times. They are the only Cenozoic rocks preserved in the area, and they record the crustal extension and basic magmatism that resulted in the opening of the North Atlantic Ocean.

Plate 17 Thin section microphotographs of olivine dolerite dykes of probable early Tertiary age from the north-west coast of Bardsey Island [1152 2226]. Euhedral laths of fresh plagioclase feldspar are partially enclosed by augitic clinopyroxene. Olivine crystals (centre left) are strongly altered. A – plane polarised light; B – crossed nicols. Field of view = 1.5 mm side to side.

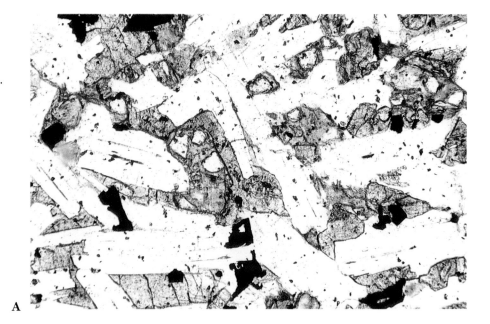

FIVE
Structural geology

There is a marked contrast between the complex structural geology of the Monian basement rocks and the less complicated structure of the overlying Ordovician strata. The Monian rocks were deformed prior to the deposition of the Ordovician rocks, which are only locally strongly cleaved and folded, although they have been tilted *en masse* towards the south-east. Only the earliest events affecting the Monian rocks are unequivocally pre-Arenig in age. The less complex deformation history of the Ordovician sequence is interpreted as due in part to the protective effect of the massive Sarn Complex basement. The most prominent structural feature of the Monian basement is the Llŷn Shear Zone, a narrow, steep belt of pre-Arenig, semischistose mylonitic rocks that has provided a focus for later brittle fault movement affecting both basement and Ordovician cover (Gibbons, 1989). The most deformed Ordovician strata are typically associated with fault zones.

STRUCTURES IN THE PRE-ORDOVICIAN ROCKS

The earliest penetrative fabric recorded by the Gwna Mélange produced an anastomosing slaty cleavage (S_1) within the mélange matrix. This fabric encloses and, in places, partly penetrates the mélange clasts. Within clasts where bedding is still preserved, an early bedding-parallel foliation is commonly discernable and was imposed prior to, or during mélange development. Within hard clasts such as the microgabbros at Porth Widlin and the granites on Bardsey Island, the earliest deformation is expressed as cataclastic and protomylonitic textures, indicating that the mélange has undergone considerable shearing. It remains unproven whether the imposition of shearing deformation postdates mélange development or occurred as the mélange was being produced. Nevertheless, in the field it is hard to separate cataclastic effects within the mélange from the disruption that produced it, and so the development of the early fabric is deduced to have occurred during the initial disruption, i.e. the mélange was produced by shearing deformation operating on variably lithified rocks at relatively shallow levels in the crust. This conclusion is generally contrary to that of Schuster (1980) who claimed a general lack of penetrative deformation within the matrix and who, with strong evidence of soft sediment deformation, believed the entire pattern of block and matrix to be the result of an essentially sedimentary process.

Only rarely are early folds (F_1) seen in the Gwna Mélange; they occur within bedded clasts. The best examples are to be seen along the south-west coast 300 m south-west of Braich y Pwll [1353 2552] (Plate 18), where they form a prominent packet of recumbent, tight to subisoclinal folds. These folds are interpreted as having developed synchronously with the surrounding S_1 fabric.

The early cleavage in the mélange is folded into a series of large, south-easterly verging F_2 antiforms and synforms, the axes of which plunge mostly south-west on the mainland (Figure 4) and north-east on Bardsey Island. Because of the south-easterly vergence of these folds, the south-east-dipping fold limbs are generally steeper (and locally overturned) than those dipping north-west. A steeply north-west-dipping S_2 cleavage is commonly axial planar to these folds, although this is usually not as well

Plate 18 Recumbent F_1 folds in 'Gwyddel Beds' (a large, bedded clast within the Gwna Mélange) on the coast [1353 2552] south of Braich y Pwll.

Plate 19 Upright F_2 folds in clast of siltstones and sandstone ('Gwyddel Beds'), which plunge gently north-east on coast [1355 2555] south of Braich y Pwll. View to north-east.

developed as the earlier foliation. These folds are exposed along the south-west coast from Braich y Pwll [1357 2590] to Trwyn Bychestyn [1510 2420] (Plate 19). At Braich y Pwll, typical chaotic mélange is exposed in the core of a south-west-plunging antiform that extends inland for at least a kilometre (Figure 4). A steep, south-east-dipping F_2 fold limb was traced along the coast for 300 m to a complementary synform [1353 2553]. South-eastwards along the coast, the same folding style may be traced in the mélange, the best example of these folds forming an antiform at Trwyn Maen Melyn [1380 2518].

Along the north-west coast, especially north-east of Porth Oer (Whistling Sands) [1670 3000], the S_1 foliation in the mélange, and the corresponding alignment of clasts, is generally very steeply dipping, and probably lies on the steep limb of a south-easterly verging F_2 fold. On such a steep limb the early S_1 foliation and the north-west-dipping cleavage (S_2) are practically parallel. Examples of smaller-scale south-easterly verging folds are exposed at several localities along this coastline, e.g. Porth Colmon [1965 3423], Porth Wen Bâch [1897 3378], Porth Llong [1665 3087] and around Porth Gwylan [2180 3710; 2128 3645]. The geometry of Monian D_2 structures is similar to that of Ordovician D_1 structures, and it is considered likely that both sets of structures developed in response to the same Acadian (end-Caledonian) deformation event.

There are three later phases of minor folding (F_3–F_5) in the Gwna Mélange. None of these deformation phases has produced any large-scale folds, but all of them locally produce abundant minor structures. The earliest (F_3) of these produced folds with a moderate to gently dipping axial surface. They are most common along the north-west coast where the axes generally plunge at low angles to the south-west. In some areas the folds are associated with the development of a weak, north-west-dipping, axial, planar, spaced cleavage, e.g. on Bardsey Island [1256 2222]. Good exposures of these folds also occur south-west of Porth Ysgaden [2184 3724], in Porth Tŷ-mawr [1890 3310] and at Traeth Penllech [2070 3490 to 2030 3435].

A later deformation phase (D_4) has locally produced gently ENE-plunging open monoclinal folds. A crenulation lineation and conjugate kink folds are also associated with this deformation. The best example is in Porth Ysgaden [2225 3746], where they locally warp the moderately northerly dip into a subhorizontal orientation. Elsewhere, they are generally much less obvious, but nevertheless can be responsible for locally anomalous variations in dip.

The final fold phase recognised within the Gwna Mélange produced a series of open, north-west- (or NNW-) to south-east- (or SSE-) trending cross folds (F_5) with steep axial surfaces. The plunge of the F_5 fold axes is strongly influenced by the previously existing dip of the rocks and is generally fairly steep (45° to 65°) and towards the north-west (or NNW). These cross folds typically occur in isolated groups; there are good examples on Bardsey Island, where six folds run out to sea obliquely to the western coastline. The best example of these folds runs across the beach in the extreme north-west of the island [1148 2248] where several conspicuous clasts of limestone, quartzite and granite are folded about a moderately NNW-plunging axis. On the mainland the best example of cross-folding occurs on the coast at Trwyn Gareg-lwyd [1652 3189], where all earlier structures are affected. Although produced by a relatively minor tectonic event, these F_5 folds have had a significant effect upon outcrop pattern by locally altering the overall strike direction from north-east towards east–west. The folding at Trwyn Gareg-lwyd induces a pronounced swing in strike direction that is in turn picked out by a dramatic change in coastal geomorphology; to the south, the coast is at right angles to the structures and has a rugged, indented form that contrasts with the more typically smooth, strike-parallel morphology of the north-western coast on Llŷn.

The structure of the Llŷn Shear Zone is dominated by the subvertical mylonitic fabric discussed in chapter two, which is interpreted as gradational into the S_1 fabric in the Gwna Mélange; the clearest evidence for this is shown on the coastline west of Trwyn Bychestyn [1490 2450] (Matley, 1928; Shackleton, 1956; Gibbons, 1983). As in the Gwna Mélange, the S_1 fabric in the Llŷn Shear Zone is contorted by many minor folds. The earliest of these folds are asymmetric towards the south-east, possess a steep north-west-dipping cleavage and are correlated with the F_2 folds exposed so commonly within the Gwna Mélange. The best examples of these folds are in the coastal section on Mynydd Bychestyn [1520 2420] and inland on Mynydd Ystum [1890 2855]. The F_3 folds in the Gwna Mélange were also recognised in the Llŷn Shear Zone on Mynydd Bychestyn where they display gently north-westerly dipping axial surfaces. They are especially common along the east side of this headland where they appear to be associated with the thrusting of these rocks over the Ordovician sequence. The later minor fold phases (F_4 and F_5) seen in the mélange have not been recognised in the Llŷn Shear Zone.

The Sarn Complex does not record the polyphase folding seen in the mélange and the schists along its north-west margin, although adjacent to the Llŷn Shear Zone it has developed a strong mylonitic fabric. This massive block of coarse crystalline rock does not possess the structural anisotropy suitable for the development of folds and is assumed to have acted as a resistant block against which the foliated Monian rocks to the north-west were deformed. In several places, the Sarn Complex rocks contain foliated amphibolitic lithologies, e.g. at Meillionydd [2185 2920] and near Crugan Bâch [2125 2997]. These rocks are interpreted as fragments either of pre-existing basic gneisses that have been incorporated into the plutonic rocks of the Sarn Complex, or of deformed, amphibolitised, early mafic intrusions, which were later incorporated within the Sarn Complex granites.

DETAILS

The north-west coast: Porth Ysgaden to Porthorion

Excellent exposures of folded Gwna Mélange occur along this coastal stretch, with virtually continuous exposure being interrupted only by thick drift sequences at Traeth Penllech [2055 3460], Porth Iago [1675 3163] and Porth Oer [1675 3000]. Around and to the south of Porth Ysgaden, there are good exposures of slaty Gwna Mélange containing long slabs of bedded sandstone and displaying many F_3 and F_4 folds. A generally steep to moderate north-westerly cleavage dip is locally reduced to shallow or even subhorizontal values by F_4 monoclinal folds. Good examples of F_3 and F_4 structures are exposed around [2180 3720], to the south of which F_4 structures become unimportant and the north-west sheet dip resumes the steep values typical of most exposures along the north-west coast. F_3 folds, with gently north-west-dipping axial surfaces, and gentle plunges to both north-east and south-west, remain common. F_2 folds are common [e.g. 2178 3710], but many have been modified by the third fold phase to produce flatter axial surfaces that may be difficult to distinguish from F_3 structures.

The Gwna Mélange retains a similar structural style from Porth Ysgaden to Porth Colmon. A steep, north-west-dipping cleavage prevails, with common south-easterly directed minor folds (F_2 and especially F_3). The influence of rare, gently NNE-plunging F_4 monoclines is minor. F_5 kink folds and larger cross folds all plunge steeply north-west down the regional cleavage dip and locally warp the north-east strike towards a north–south direction [e.g. 1970 3422]. On and below the ice-polished wave-cut platform at Porth Colmon [1940 3435], the mélange continues to display a steep north-west cleavage dip (S_1 and S_2). Virtually all clasts are strongly flattened within the main (S_1–S_2) foliation, although some of the larger white quartzite clasts define isolated F_2 fold hinges. Farther south-west, around Trwyn Cam and Porth Llydan [1913 3420], large slabs of sandstone preserving bedding lie within the mélange. South-west-plunging F_3 folds are locally common within these slabs.

Within the Gwna Mélange at Porth Llyfesig [1860 3250], an enormous clast of purple basic lava locally includes pale green microgabbro (at Porth Widlin [1830 3246] and inland at Pen-yr-Orsedd [1915 3170]). The basic lava and microgabbro clast at Porth Llyfesig displays only a poor record of the deformation history seen in the surrounding mélange, and is strongly fractured. The microgabbro has, in particular, been responsive to shearing deformation and locally takes on a mylonitic appearance. Beyond Porth Widlin, flattened slivers of lava and lava-limestone beccias are mixed with sandstone and limestone clasts within the slaty mélange. A steep S_1/S_2 fabric continues to dominate the sheet dip, with flat F_3 folds common within slices of psammitic sediment; a good example of F_2 and F_3 minor structures is exposed in the gully at [1757 3225]. F_4 folds were not identified, but gentle F_5 cross-folds locally warp the strike. At Trwyn Gareg-lwyd [1655 3190] is the prominent north-west-plunging F_5 synform mentioned previously, with the rocks south of here displaying a steep northerly dip for over 1 km. On the north side of Porth y Wrâch [1670 3070], this northerly dip gradually curves back to the more typical north-westerly direction.

All exposures along the north-west coast from the northern margin of the district to Trwyn Glas [1645 3095] lie on a steep F_2 fold limb, with evidence for younging directions of large clasts being towards the south, for example on Trwyn Glas. Beyond this latter locality, however, is the first of a number of poorly defined south-west plunging F_2 folds. These F_2 folds dominate the Gwna Mélange structure down to the south-western tip of the peninsula.

The Gwna Mélange south-west of Porth Oer shows a dominantly steep north- to north-west-dipping fabric, and comprises mostly disrupted pillow lavas mixed with smaller amounts of limestone, red shaly mudstone, sandstone, 'Gwyddel Beds' and slaty, green, semipelitic mélange matrix. The rocks are cut by thrusts and steep faults, and F_2 fold hinges are locally present, the best developed being a synform just east of Dinas Fawr [1565 2905] and an antiform north of Porthorion [1568 2875].

Mynydd Anelog

The prominent hill of Mynydd Anelog provides one of the few areas of extensive inland exposure within the district. The northern side of the hill exposes scattered clasts of cherty sediments (Matley's 'Gwyddel Beds') mixed with various sandstones, all lying within the green slaty matrix typical of the Gwna Mélange. The southern half of the hill is mostly a large clast of cherty 'Gwyddel Beds', with a zone of red slaty mudstones, strongly folded by F_2; it is exposed in the south-east corner around Uwchmynydd [1550 2633]. Steep dips prevail over the whole area, as most of the outcrop lies upon a steep F_2 fold limb. F_2 minor fold axes maintain a consistent WSW plunge across the hill, except where late F_5 cross folds have warped them into an east–west trend. By contrast, the bedding (S_0) and bedding-parallel (S_1) fabric show much greater variation, an effect produced by a combination of F_2 folding and the deflection around the larger clasts of 'Gwyddel Beds' (especially the southern mass). The effects of both D_3 and D_4 are not apparent across the hill, although several examples of steep north-westerly plunging F_5 folds occur, the best examples of these being exposed in the fields west of Bryn-mawr [1560 2800].

In contrast with much of the hillside, the precipitous coastline provides a fine section through the sequence. Most of the cliffs display steeply dipping mélange and 'Gwyddel Beds', although south of Braich Anelog [147 276 to 147 272] a zone of variable dips, mostly gentle, marks the position of a wide asymmetric F_2 fold. This disturbance is responsible for the downfolding of the southern 'Gwyddel Beds' block to form the detached mass south-east of Braich Anelog. The fold is present around the hill to where it is displaced by a sinistral tear fault, which moves the shallow limb back out to the coastline.

The south-west coast: Mynydd Mawr to Pared Llech-ymenyn

Upright F_2 folds dominate the structure of the Gwna Mélange in the extreme south-western tip of Llŷn (Figure 4). F_3–F_5 structures are much less commonly developed than along the north-western coastline. The sheet dip throughout the south-west coast section is controlled by a series of large-scale F_2 antiforms and synforms (with steeply north-west-dipping axial surfaces) that fold an earlier, low angle S_1 fabric.

At Braich y Pwll [1355 2588], typical chaotic Gwna Mélange is exposed in the core of a south-westerly plunging F_2 anticline. The steep south-easterly limb of this fold brings down a thick, internally coherent sequence of 'Gwyddel Beds' containing two horizons of red shaly mudstones, one at the base and one approximately 15 m above this. Disrupted clasts of the basal red mudstone layer have also been incorporated within the mélange. The steep F_2 fold limb displays many south-west-plunging minor folds and a steeply dipping S_2 cleavage that cuts subvertical bedding. The beds in this steep fold limb are exposed for some 300 m southwards along the coast to a complementary synclinal hinge at the north-eastern end of a prominent inlet [1353 2553]. The plunge of this F_2 fold reverses from south-west to north-east. Beyond the synclinal fold hinge the bedding and S_1 fabric dip gently to moderately north-west, with gently north-easterly plunging minor folds. A few metres south of the synclinal hinge there is the series of previously mentioned, well-developed, recumbent F_1 folds (Plate 19) that fold bedding and are themselves modified by upright F_2 folds.

The shallow north-westerly dipping F_2 fold limb continues for some 400 m along strike to an antiformal hinge on the west side of Maen Melyn Lleyn [1384 2518]. As on the west coast of Mynydd Anelog, the dips in this area are highly variable. Although the general dip is shallow and north-westerly, the rocks are corrugated by many parasitic F_2 folds. The overall dip to the north-west has the effect of bringing the mélange back up to sea level at [1372 2543]. The actual contact between the 'Gwyddel Beds' and the slaty mélange is faulted.

The S_1 fabric in the cliffs west of Trwyn Maen Melyn is folded over a gently north-east-plunging F_2 antiform. The cliffs on the shallow, north-western limb of this antiform provide the best known exposures of Gwna Mélange, with large disrupted blocks of white quartzite defining the gentle north-westerly dip (Plate 1). Just beyond these prominent clasts, the S_1 mélange fabric curves over the antiformal fold hinge to bring the disrupted quartzite layer back down into the cliffline and produce a steep south-easterly dip. This steep F_2 fold limb was traced around the west side of the cove, east of Trwyn Maen Melyn [1397 2518], into a zone of tight folding. At the back of the cove, several faults cut out an F_2 synformal hinge. Immediately west of here is the great clast of 'Gwyddel Beds' that forms Mynydd y Gwyddel. The beds initially dip moderately north-west on the F_2 shallow limb but, at the headland [1395 2505], a moderately south-west-plunging antiform folds a complicated disrupted mélange of 'Gwyddel Beds', red slaty mudstones, basaltic lava and sandstones in a green semipelitic matrix. The rest of Mynydd y Gwyddel is composed entirely of pale weathering, cherty 'Gwyddel Beds' (Matley's type locality for this lithology). Along the coast, the steep south-easterly dip continues for some 300 m to the headland of Trwyn y Gwyddel [1420 2475] where a south-westerly plunging synform produces a southerly dip in its hinge, and a subsequent north-westerly dip on the opposite limb.

The above description emphasises how the coastline between Braich y Pwll and Trwyn y Gwyddel exposes a series of large F_2 folds of consistent NE–SW axial trend but variable plunge, and it is possible to trace most of these folds inland to the north-east. The antiform at Braich y Pwll initially runs out to sea, but intersects the coast again for 400 m at [1386 2607], and is probably the same fold seen on the south-west flank of Mynydd Anelog [1470 2680]. The complementary syncline within the 'Gwyddel Beds', where the F_1 folds are exposed, appears to run through the centre of Mynydd Mawr. The exposures on the north-west side of the hill belong to the steep F_2 limb, whereas those on the south-east side of the hill are dominated by gentle dips. This Mynydd Mawr Syncline is probably partly responsible for the strongly asymmetric geomorphology of the hill. The other main F_2 folds, that at Trwyn Maen Melyn and those on Mynydd y Gwyddel, are less easy to trace inland. This is due to poor exposure, the masking effect of the S_2 foliation (many inland outcrops are dominated by this fabric) and the fact that the whole area consists of an interwoven series of variably plunging F_2 folds.

At Porth Felen [1438 2498], the west side of which exposes the red mudstones at the base of the 'Gwyddel Beds', a series of steep faults at the back of the bay brings a mélange rich in pillow lavas against the 'Gwyddel Beds' of Mynydd y Gwyddel. The slaty S_1 fabric within this mélange dips moderately north-west and, in places, one may discern the steep cross-cutting S_2 cleavage (Plate 20) which shows the rocks to lie on a shallow F_2 fold limb. Several north-west-dipping thrusts slice through this mélange, with the most prominent example forming a zone immediately above a conspicuous limestone horizon at [1456 2473]. These thrusts display strong mineral fibre lineations indicating a south-easterly shear direction. They are interpreted to be of Acadian age and were probably produced during the D_3 event in the Monian deformation history. The previously mentioned thrust zone at [1456 2473] shows a sliver of cleaved grey mudstone and siltstone in the hanging wall immediately above the most prominent thrust plane. Although undated, these deformed sediments show a greater lithological similarity to the Ordovician sequence than to the surrounding Monian rocks. It is considered likely, therefore, that this thrust has sliced through a downfaulted block of Ordovician sediment which originally lay to the north-west, incorporating a sliver of this sediment within the thrust zone.

At Pared Llech-ymenyn the gently north-west-dipping Gwna Mélange lies in the hanging wall of a thrust that overrides a dolerite intrusion and steeply dipping, highly deformed metasediments. These metasediments include lenses of limestone, white quartzite and basalt, and are interpreted as more highly deformed Gwna Mélange. They become increasingly deformed and metamorphosed to phyllitic semischists south-eastwards into the Llŷn Shear Zone, exposed on the headland of Trwyn Bychestyn (p.23). The thrusts that cut the Monian rocks along this coastal section are interpreted to be part of the post-Arenig Parwyd Thrust system (see below).

Plate 20 Clasts of sandstone within Gwna Mélange exposed on cliffs east of Porth Felen [1445 2490]. The semipelitic matrix to the mélange displays a steep early cleavage cut by a spaced second fabric dipping more gently towards the north-west. Looking north-east.

Bardsey Island (Ynys Enlli)

The Gwna Mélange of Bardsey Island displays a well-developed S_1 foliation which dips north-west over most of the island, except in the southern isthmus where dips are mostly 30° to 50° to the north.

South-easterly verging F_2 folds cross the east side of the island and show a north-easterly plunge, so that higher structural levels in the mélange are seen from south-west to north-east across Mynydd Enlli [1230 2175]. One of the most prominent of these F_2 folds is a north-easterly plunging, much faulted synform [1198 2094] west of Pen Cristin that runs into Mynydd Enlli. A minor antiformal disturbance is traceable along much of the summit ridge of Mynydd Enlli, the almost vertical steep limb enhancing the precipitous nature of the eastern slope in a manner analogous to (but in the opposite sense) the western slopes of Mynydd Mawr. Similar flexures run out to sea on the north coast. These folds possess a generally weak, steeply north-west-dipping, axial planar fabric. The style and orientation of these structures shows that they are probably the equivalents of the F_2 structures seen on the mainland to the north-east. Unlike the mainland exposures, however, F_2 minor folds are relatively rare on Bardsey Island. On the western coast, the S_1 fabric has a generally moderate to steep dip, and relatively rare F_2 minor folds indicate that the exposures lie on a steep overturned F_2 limb. An F_2 synformal hinge between these exposures and the eastern side of Mynydd Enlli appears to have been faulted out.

Rare, flat-lying minor F_3 folds show axial surfaces dipping gently north-west or north; they commonly lie above low-angle faults, suggesting that they were produced during thrusting. The rocks are also folded by gently ENE-plunging F_4 monoclines. The strongest development of these F_4 structures is on the west coast at Porth Hadog [1127 2096], where they warp the moderate northerly dip into a subhorizontal orientation. Elsewhere, they are generally less obvious, but are responsible for much of the variation in dip along the west coast.

An unusual feature of the structure on Bardsey Island is the presence of six prominent NNW- or north-west-plunging F_5 fold zones. The best exposed zone crosses the beach in the extreme north-west [1147 2247] where several conspicuous clasts of limestone, quartzite and granite are folded about an axis that plunges at around 44° to the north-north-west. The angle of plunge of these cross-folds is strongly influenced by the orientation of the previously existing sheet dip; where a steep foliation dip prevails, as on the hillside above the Methodist Chapel [1218 2215], the F_5 plunges are correspondingly steep.

POST-ORDOVICIAN STRUCTURES

Folds

In general the Ordovician sediments in the Aberdaron area are not strongly folded and maintain a consistent dip of around 30° to 50° towards the south-east. The best examples of folding are to be seen along the coast south of Porth Simdde [1665 2604], south of Porth Meudwy [1635 2518], against the Wig Fault at Wig [1861 2574], in Porth Cadlan [2001 2616] and in the stream north of Porth Ysgo. These folds belong either to a north-east-plunging set associated with south-easterly directed compression (Trwyn Cam Syncline and Porth Cloch Anticline), or to a younger, upright set of variable plunge (the F_3 group of Hawkins, 1983). These younger folds are of sporadic distribution, and are commonly associated with faults, as at Wig.

Several previous authors have interpreted the Ordovician outcrop immediately west of the Daron Fault as defining a south-west-plunging 'Aberdaron Syncline'. Matley (1928) first referred to this structure, basing it upon anomalous west-north-westerly dips in the rocks around Bodwrdda. Shackleton (1956) also described this structure, terminating its eastern limb against the Daron Fault. Hawkins (1983) later cast doubt on the evidence for this fold. Exposure in this area is extremely poor, except in Afon Daron where, although bedding direction is somewhat variable (as noted by Hawkins), there is a definite tendency towards a south-south-westerly dip, for example in the stream bed [1907 2742] and along the west bank [1919 2769]. Despite statements to the contrary

(Hawkins, 1983), the Pen y Cil Sill can be traced as far as the Afon Daron, although there is no evidence for the sill defining a synclinal hinge. Instead, the anomalous dips in the Afon Daron exposures are attributed to the effect of faulting within the Daron Fault system.

Two prominent fold zones occur along the coast south-west of Aberdaron beach. These folds are the Trwyn Cam Syncline and the Porth Cloch Anticline. The Trwyn Cam syncline is exposed along the west side of Aberdaron Bay north of Porth Meudwy [1665 2603], where it is accessible only at low tide but visible from Aberdaron beach (Matley, 1932). It is a broad, open, synclinal fold that forms a footwall syncline to an overriding, north-west-dipping, steep thrust. The syncline plunges at 16° towards N060°. The south-eastern limb of this fold curves into a hanging wall anticline above another moderately north-west-dipping reverse fault, which emplaces laminated siltstones upon the Meudwy Sandstone Member at Trwyn Cam [1665 2602].

The Porth Cloch Anticline is a previously unrecorded structure and is exposed along a stretch of coastline 150 m north of Porth Cloch [1635 2518]. Most of this coastline is occupied by the Pen y Cil Sill, but at this locality a combination of folding and faulting exposes a small inlier of Ordovician sediments beneath the sill. The main structure is a thrust-related south-easterly verging asymmetric anticline which plunges at about 40° towards N035°. Well-developed axial planar cleavage dips about 54° to the north-west. Minor duplex development within the core of the fold indicates that it is a tip-line structure in front of a propagating thrust. This structure may represent the deformation front of the Parwyd Fault system and shows that thrust-related deformation occurs within the Ordovician sequence over 500 m from the Precambrian–Ordovician contact.

It is difficult unequivocally to correlate deformation phases between the Monian basement and Ordovician cover. However, the main south-easterly verging folding and cleavage-forming event in the Ordovician cover is similar in style to the south-easterly directed deformation that produced the F_2 folds and F_3 folds and thrusts in the Gwna Mélange and the Llŷn Shear Zone. Fragments of mylonitic lithologies in the basal Ordovician strata, such as those exposed above the unconformity at Wig, prove the earliest deformation in the basement (i.e. that which produced the Llŷn Shear Zone mylonites and the S_1 fabric in the Gwna Mélange) to be pre-Arenig. A similar association occurs in Anglesey, where fragments of foliated Monian lithologies are common in the Ordovician cover. The simplest interpretation of these data is that all post-D_1 folding and cleavage formation within the Gwna Mélange and the Llŷn Shear Zone is post-Llanvirn and, by analogy with areas elsewhere in Wales, probably of Acadian (Devonian) age.

Faults

Four main post-Arenig brittle fault systems occur within the district and strongly influence the outcrop pattern of the map. One of these, the Parwyd fault system, forms the north-east-trending faulted junction between Monian and Ordovician rocks, which runs from Parwyd to Mynydd Ystum. The second system forms a series of steep, north-west-trending cross faults, such as the Afon Saint Fault. The third fault set is again steeply inclined, but trends NNE–SSW and includes the Daron Fault, which separates the two Ordovician successions. Finally, there is a complex fault system associated with the manganese mines in the Nant y Gadwen and Benallt areas in the south-east corner of the district. Each of these fault systems is discussed below.

Parwyd fault system

The relationship between the Precambrian and Ordovician rocks west of the Daron Fault has been a subject of differing interpretations (chapter one). At the coast, mylonitic metasediments in the Llŷn Shear Zone can be seen to have been thrust over Whitlandian marine sediments on the west side of Parwyd [1534 2435]. The rocks in the footwall are cut by several faults, which link into the south-easterly directed thrust system. Despite this south-easterly movement the fault at Parwyd dips 40° to 50° to the south-west and is interpreted as a sidewall ramp in the thrust system. The Monian–Ordovician boundary is extrapolated inland north-east to Ty-tan-yr-allt [1663 2625], where Matley (1939) recorded an exposure of a low-angle north-west-dipping thrust with Precambrian rocks in the hanging wall. Farther north-east, this compressional fault contact is offset by the Afon Saint Fault (see below), beyond which there is no topographical expression, but the approximate position of the boundary may be continued towards Mynydd Ystum.

In a track [1807 2808] north of Hendre-Uchaf, finely schistose mylonitic rocks of the Llŷn Shear Zone crop out within a few metres of Ordovician siltstones, and the nature of the contact, whether unconformity or fault, has been the subject of some debate. A trench was excavated to a maximum depth of 2 m parallel to and immediately east of the track. This excavation proved that the contact between these two rock units is marked by a steep south to south-west-dipping fault along which there is 50 mm of brown, limonitic clay gouge. There is no unconformity here and the Ordovician mudstones terminate abruptly against the fault.

Immediately to the south-west edge of Mynydd Ystum [1851 2830], both Precambrian and Ordovician rocks are again in close proximity. Two trenches were dug in this area, the first running south-west from [1849 2831] and the second running south-east across the Precambrian–Ordovician boundary at [1846 2828]. The first trench proved the presence of a vertical fault which strikes NW–SE, forming the feature which terminates the south-western end of Mynydd Ystum. The second trench also proved a fault contact between Ordovician siltstones and Monian mylonitic metasediments. It trends north-east and dips steeply north-west; it has a 0.2 m-thick zone of brown limonitic clay gouge and is mineralised with oxidised pyrite. This fault is probably cut by the north-west-striking fault a few metres to the north-east at [1849 2829].

The contact between the Precambrian and Ordovician west of the Daron Fault, except beneath the Parwyd outlier, is in all cases a fault. This junction is a low-angle thrust in the south-west, and a higher angle reverse fault in the north-east of the district. This boundary fault, named the Parwyd Fault on the map, has emplaced the Precambrian basement upon Whitlandian sedimentary rocks. The south-east-directed compression responsible for the generation of this fault system also produced the folding and main cleavage seen in the western part of the Ordovician outcrop.

Exposure of the Ordovician sediments on the coast south-west of Aberdaron reveals many faults, most of which are high-angle reverse faults or lower-angle thrusts associated with the south-east-directed deformation. Good examples of compressional faulting, folding and cleavage development occur at Trwyn Cam [1665 2604] and in the cliffs between Porth Meudwy and Porth Cloch [1636 2521]. Faulting is particularly common around Parwyd where, south-east from the Parwyd Thrust, a steep fault has brought the Parwyd Gneisses against Arenig mudstones [1549 2416]. Judging from the coastal exposure, it is likely that the unexposed ground inland hides many faults within the Ordovician sequence. Similarly, steep north-east-striking faults are common in the Gwna Mélange exposed on the north-west coast and are probably common inland beneath the drift.

North-westerly striking faults

A conspicuous set of steep faults, comprising the Afon Saint, Aberdaron and Ystum faults, displaces the outcrop pattern in the area west of the Daron Fault. The Afon Saint Fault is marked by a valley, and the fault plane has been worked for baryte (chapter six). It offsets the Monian–Ordovician boundary by 170 m and displays sinistral movement. It is correlated with the steep fault at Ogof Pren-côch, on the north-west coast [1542 2637]. The Aberdaron Fault also offsets the Precambrian–Ordovician boundary in a sinistral sense by about 430 m. The trace of this fault is less well constrained : to the south-east it is lost beneath thick drift at Aberdaron, and to the north-west it is extrapolated to form the south-western termination of Mynydd Carreg. The Ystum Fault, exposed by trenching as described above, forms the scarp boundary at the south-western end of Mynydd Ystum [1835 2840].

Other north-westerly to NNW-trending faults occur in several locations, such as around Porth Meudwy [1635 2558], on the west side of Parwyd [1535 2441] and on Mynydd Anelog [1525 2800]. It is likely that these faults have been produced by continuing south-eastward-directed compression and that at least some of them may represent compartmental faults synchronous with the thrusting seen at Parwyd.

Daron Fault

The steep Daron Fault forms a conspicuous topographical feature running inland from the coast at Wig. It was proved in the Afon Daron gorge [1912 2745; 1920 2767], but is lost below drift farther north. It brings Monian mylonites against Ordovician sedimentary rocks and is likely to represent the brittle reactivation of a more fundamental basement shear zone. The Ordovician mudstones exposed close to the fault in the Afon Daron gorge have a strong NNW-striking cleavage, interpreted as having formed during faulting. Similarly, the Ordovician sediments exposed on Trwyn y Penrhyn [1852 2465] show a steep NNW- to NNE-striking cleavage, which increases in intensity towards the fault.

The Wig Fault, between Monian mylonites and Ordovician rocks on the coast at Wig [1858 2582; 1858 2574], is probably a splay from the main Daron Fault. It dips at around 80° to the west and exhibits a well-developed cataclasite, derived from the fragmentation of both Precambrian and Ordovician lithologies. Most mineral fibre lineations in the Monian rocks close to the main Wig Fault plunge gently south-westward and suggest oblique strike-slip movement with a downthrow towards the south. This interpretation is supported by the fact that the unconformity surface at Wig is downthrown some 70 m from its position at Penrhyn Mawr Farm on the east side of the Wig Fault.

The fault at Ogof Lleuddad [1907 2554] runs parallel to the Wig Fault and, like the latter, shows associated compressional folding. However, the same laminated siltstones were found on both sides of this fault and there is no evidence for any major movement across it. It is likely that this fault at Ogof Lleuddad represents a minor splay from the main Daron Fault.

Faults in the manganese mining belt

The faults within the manganese mining belt in the south-eastern corner of the district are poorly exposed, but old mine records and magnetic surveying (Groves, 1947; Brown and Evans, 1989) have enabled their position to be traced. Matley (1932) inferred the presence of a major fault, trending ENE approximately along the line of the Rhiw–Aberdaron road, displacing the Nant portion of the mining belt from the Benallt portion. The results of the recent magnetometer survey (Brown and Evans, 1989) support the presence of such a fault, but show that it produces considerably less displacement than suggested by Matley. Faults were proved in the glory hole at Nant Mine [2118 2672] and in the opencast workings at Benallt Mine [2219 2828]. The prominent fault in Nant y Gadwen appears to be a NNW–SSE-striking structure that separates the south-east-dipping succession seen on both east and west sides of the valley from south- to south-west-dipping strata in the north-east part of the valley (p.40). These anomalous south-westerly dips are interpreted as being due to compressional folding on the eastern side of a fault running north-east towards Benallt. Anomalous dips of this nature are similar to those associated with folding on the east side of faults above Porth Ysgo [2086 2661] and in the Afon Daron gorge [1908 2744]. At Benallt Mine, a 5 m-wide, subvertical fault zone is associated with the development of steep cleavage and the generation of tight, southerly plunging folds within Ordovician strata. A pair of east–west faults displace an approximately 100 m section of the Footwall Sill across the manganese mudstones, as proved by a trial adit (Groves, 1947) and magnetic data (Brown and Evans, 1989).

SIX

Quaternary

Interpretation of the Quaternary 'drift' deposits of the district has taken into account both geological observations on the sediments and geomorphological evidence from associated landforms.

The evolution of the present landscape began long before the Quaternary (last 2.4 million years), and some discussion of the development of the preglacial topography is essential if the influence of the ice sheets is to be fully assessed. Such an assessment also relies on the erosional evidence, including that of striations, which indicate the direction of ice movement, and features produced by subglacial meltwater, which provide some evidence of conditions under the advancing ice. Quaternary depositional environments are reflected by a range of lithological facies, including glacial melt-out and flow tills, and gravels deposited in englacial or braided outwash streams.

Coastal sections and rare inland exposures in glacial deposits show complex and rapid lateral and vertical facies changes which are not recorded as surface features. All of the glacial deposits are, therefore, included in a single lithological category as 'glacial deposits undifferentiated'. However, subdivision of the drift is possible on the basis of topography, providing a useful indicator of probable thickness and an aid to genetic interpretation.

Postglacial drift deposits, including head and alluvium, are of restricted occurrence in the district.

PREVIOUS WORK

Although some earlier workers mention briefly the drift deposits of western Llŷn (Trimmer, 1841; Ramsay, 1881; Lewis, 1894), the first detailed work was by Jehu (1909), who recognised a preglacial wave-cut platform and described drift sections from Porth Oer, Porth Meudwy, Porth Pistyll and Aberdaron Bay. He proposed a tripartite division of the drift into upper and lower boulder clays separated by intermediate sands and gravels of probable interglacial status.

The significance of the raised beach platform was discussed by Whittow (1960), who attempted to correlate it with similar features in southern Britain and eastern Ireland. Such correlations are now considered to be of little value, since the height of platforms is not necessarily related directly to any single past sea level, and deposits lying on them need not immediately postdate marine planation.

Synge (1964) visited a number of sites on Llŷn, but only described one section within the district. At Porth Simdde he recognised a purple calcareous till resting upon coarse angular material, and suggested that the direction of ice-flow in this part of the peninsula was from north-east to south-west.

Saunders (1968a) used analysis of till fabrics from sites throughout Llŷn, including Porth Oer and Aberdaron, to dispute the conclusions of Jehu (1909) and Synge (1964), suggesting that the dominant ice-flow direction was from north-west to south-east. In describing glacial drainage phenomena, Saunders (1968b) concentrated on features farther east, but included Porth Meudwy as an overflow channel.

The Quaternary history of the area was summarised and compared with other available evidence by Whittow and Ball (1970), who accepted the model of ice-flow from north-west to south-east. Later work (e.g. Campbell and Bowen, 1990) has relied largely upon this summary.

PREGLACIAL LANDSCAPE EVOLUTION

The subdrift bedrock topography of western Llŷn, dominated by isolated hills and broad plains (Plate 21), was considered by Brown (1960) to form part of the 'coastal plateaux' of Wales. He suggested that the isolated hills might be remnants of an upland plain rising towards Snowdonia. The rest of the landscape he attributed to marine erosion during the Pleistocene. Whittow (1957, p.8), although noting that 'a cursory examination would conclude that the landscape was merely one large subaerial peneplain with a few residual monadnocks', recognised erosional evidence for several periods of relatively high sea level.

However, the similar topography of Anglesey, with its broad plain and isolated hills, previously interpreted as the product of two episodes of Tertiary marine planation (Brown, 1952, 1960; Embleton, 1964), has been reinterpreted as the result of deep tropical weathering (Battiau-Queney, 1980, 1984, 1987). The isolated hills are seen as inselbergs, emerging from the surface of deep tropical soils. Deeply weathered bedrock recognised by Greenly (1919) on a number of hills in Anglesey is regarded as a remnant of the saprolite cover (Battiau-Queney, 1984). Similar remnants of deep weathering have been reported from Scotland (Hall, 1985, 1986, 1987) and Ireland (Battiau-Queney and Saucerott, 1985).

The similarity of the landscapes of Anglesey and western Llŷn suggests that a similar origin might be suggested for the subdrift plain and the isolated hills in the Aberdaron area. No deep weathering has been recognised on the hills, but there is one coastal site, at the northern end of Porth Oer [1676 3012], where pillow lavas in the Gwna Mélange are weathered to such a degree that they can still be cut with a spade at a depth of 2 m, although hard corestones remain throughout.

Plate 21 The rocky hill Mynydd Carreg [163 291] which rises abruptly above the plain of thick drift extending towards the coast at Porth Oer.

RAISED SHORE PLATFORM

At a number of localities around the coast, a gently sloping surface, which has been interpreted as a raised marine platform (Whittow, 1960), extends beneath a cover of drift. The platform occurs south of Porth Oer [1632 2987], where it has been dissected to produce low stacks, and is particularly well developed around Porth Colmon [1950 3430]. It is exposed on Dinas Fawr [156 291], Dinas Bâch [158 294] and on Bardsey Island from Penrhyn Gogor [116 226] to Porth Solfach [115 213], and from Henllwyn [113 209] to Pen Cristin [121 209]. At none of these localities is the notch at the junction of platform and cliff exposed, but Whittow [1960] noted its presence at about 25ft (7.6 m) at Porth Dinlleyn [2781 4192], 6.7 km north-east of the district.

The platform was recognised by Jehu (1909) and its full extent mapped by Whittow (1960), who termed it the predrift raised beach platform, or 25-foot platform. Platform remnants occur at a similar height around the coast of Anglesey (Greenly, 1919; Hopley, 1963; Whittow, 1965). Comparable remnants also occur in south-west Wales, Devon, southern and eastern Ireland, the Channel Islands and the Atlantic coast of France (Bowen, 1977; Keen, 1978; Kidson, 1977; Mottershead, 1977; Guilcher, 1969; Stephens, 1970, 1980). In Gower, a raised platform is overlain by raised beach deposits which have been assigned to the Ipswichian interglacial, Oxygen Isotope Stage 5e (Bowen et al., 1986). However, correlation of the platform remnants is difficult since the height of a platform is not necessarily directly related to any single past sea level, and deposits lying on it need not immediately postdate marine planation. The platform in Llŷn retains clear evidence of glacial abrasion; thus the only firm conclusions that can be drawn are that the platform predates the last glacial advance and that it was formed at a time when sea level was higher than at present.

EVIDENCE OF QUATERNARY GLACIATION

The evidence of glaciation in Llŷn and north-west Wales was reviewed by Whittow and Ball (1970), who considered the drift stratigraphy to have resulted from 'a complex ebb and flow of Northern and Welsh ice'. They recognise that two glacial phases (they refer to them as stages but the term is not used here because of its chronostratigraphical connotations) were responsible for the deposits of Llŷn (Figure 12). During the first phase it is suggested that Irish Sea ice crossed the peninsula from the north-north-west. At the same time, Welsh ice moved southwards into Tremadoc Bay, its western boundary impinging on St Tudwal's Peninsula. During the second (Main Anglesey Advance) phase much of Llŷn is considered to have been ice free, with ice crossing Anglesey from the north-east and entering Llŷn only between Nefyn and Porth Dinllaen. Welsh ice moved west into Tremadog and Cardigan bays, forming the celebrated sarns (Campbell and Bowen, 1990) and reaching as far as St Tudwal's Peninsula.

Although this reconstruction was based largely on morphological evidence and till stratigraphy (the 'count from the top' procedure criticised by Bowen, 1973), it has not been seriously challenged. The second phase (Main Anglesey and Arvon advances of Whittow and Ball, 1970) is retained by Bowen et al. (1986) as the 'Gwynedd Readvance' during wastage of the Dimlington Stadial (Late Devensian) ice. In this study, however, evidence of only one glacial phase has been recognised (McCarroll, 1991).

Erosional evidence

The clearest evidence of glacial erosion in the Aberdaron area is in the form of striations and grooves produced by clasts at the base of the glacier (Plate 22). The best examples occur around the coast, where glacially abraded

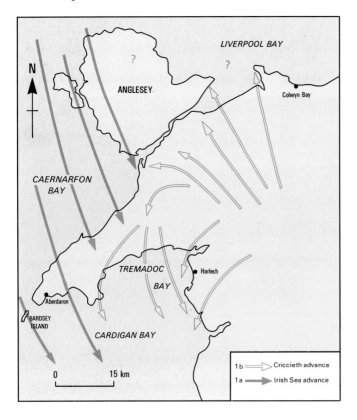

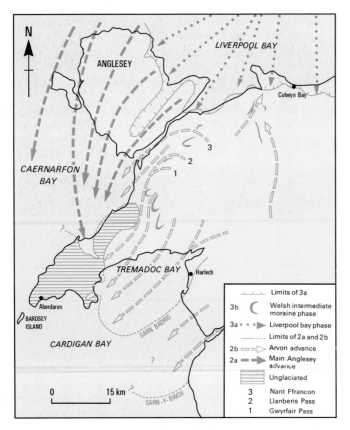

Figure 12 Glacial stages in north-west Wales as recognised by Whittow and Ball (1970). Reprinted, by permission of the publishers, from Whittow and Ball, 1970 (figs. 2.1 and 2.2).

bedrock has recently been exposed or can be exhumed from beneath a protective cover of drift. Inland, much of the exposed bedrock on the hill tops has been frost shattered, removing evidence of glacial abrasion. Elsewhere, exposed bedrock surfaces have generally been chemically weathered.

Scratches which might be confused with glacial striations can be produced by a variety of mechanisms (e.g. differential weathering of cleavage or slickensides), and care must be taken to exclude these. Good examples of artificial 'striations' produced by farm machinery occur on a track north of Porth Meudwy [1621 2562] and on bedrock outcrops in a field near Craig Fael [2151 2996]. In this study, striations were recorded only from smooth, glacially polished, fine-grained lithologies (Plate 22A); wherever possible, a fresh surface was exhumed. On rougher, coarse-grained and contorted lithologies, clear deep grooves rather than fine striations were recorded (Plate 22B).

Although striations and grooves clearly indicate abrasion by clasts travelling in a particular direction, they must be interpreted carefully in order to separate indications of the regional ice flow pattern from very local effects. Ice is capable of deforming around obstacles, and so isolated striae may provide a confusing picture of regional ice flow. Between Hen Borth and Pen y Cil [1584 2422], for example, striations on gently sloping dolerite surfaces are aligned north–south, whereas just down slope, where the surfaces are much steeper, the direction of the striations swings around to between N035° and N060°.

The striations of western Llŷn display a clear trend from north-east to south-west in the northern part of the area, to north–south alongside the hills in the south-west (Figure 13). Only one site, at Porth Solfach on Bardsey Island, suggests ice flow from the north-west. Although Saunders (1968a) recognised some of the striations recorded here, he suggested that the ice actually flowed from the north-north-west, the basal layers being deflected by the coastal fringe of Precambrian rocks to flow from the east-north-east to west-south-west. The evidence for this was based largely upon the orientation of clasts in diamicts ('tills'). However, reconstruction of ice movements from clast fabrics assumes that the diamicts were lodged or deposited by melting out in situ without redeposition as flow tills. It will be demonstrated that such an assumption is not warranted for many of the diamicts of Llŷn.

The simplest interpretation of the striations of western Llŷn is that they reflect ice movement from the north-east in the northern part of the area, veering to northerly alongside Mynydd Anelog, Mynydd Mawr and Mynydd Enlli in the south-west. With this interpretation of ice flow over western Llŷn there is no need to invoke a 'Main Anglesey Advance' (Whittow and Ball, 1970) or a 'Gwynedd Readvance' (Bowen et al., 1986) of the Irish Sea ice. The erosional and depositional evidence of both Anglesey (Greenly, 1919; Whittow and Ball, 1970; Harris, 1989) and Llŷn can be interpreted in terms of a single advance and retreat. The 'Clynnog Fawr moraine', which Synge (1964) proposed as a major ice limit, can be explained by a prolonged stillstand of active marginal ice against the hills of the north Llŷn coast (McCarroll, 1991).

Plate 22 A. Striations and crescentic fractures on quartzite in the Gwna Mélange west of Porth Colman [1938 3432]. B. Glacial grooves on the raised shore platform west of Porth Colman [1938 3432].

A

B

The large-scale landforms of western Llŷn probably cannot be taken as indicators of ice movement direction. It is tempting to interpret the elongate hills Mynydd Carreg, Mynydd Ystum and Mynydd Penarfynydd as 'crag and tail' features, with relatively resistant lithologies to the north-east protecting a tail of less resistant rock to the south-west. However, because the Caledonian structures are approximately parallel to the assumed ice flow direction, it is difficult to separate the influence of structural and lithological control. Moreover, the preservation of the striated coastal platform and of some sediments which predate the last advance argue against massive erosion by late Devensian ice. As discussed above, the major landforms are probably preglacial and have been modified only by glacial advances and periglacial conditions during the Quaternary.

The landscape of south-western Llŷn retains some landforms resulting from the erosive action of subglacial meltwater. On the west coast of Aberdaron Bay, for example, short valleys are cut through bedrock at Porth Simdde [166 265] and Porth Meudwy [162 256]. These valleys support small, underfit streams and are partly drift-filled. The simplest interpretation is that they were produced subglacially. In Uwchmynydd, the land rises towards the coast and the ice would have been forced to move over this obstruction. The preglacial drainage probably flowed parallel to the coast, from south-west to north-east, to join the headwaters of the Afon Cyll-y-Felin, or turned north-east to the sea at Porth Llanllawen. The basal meltwater, under high hydrostatic pressure, was able to breach the coastal ridge along lines

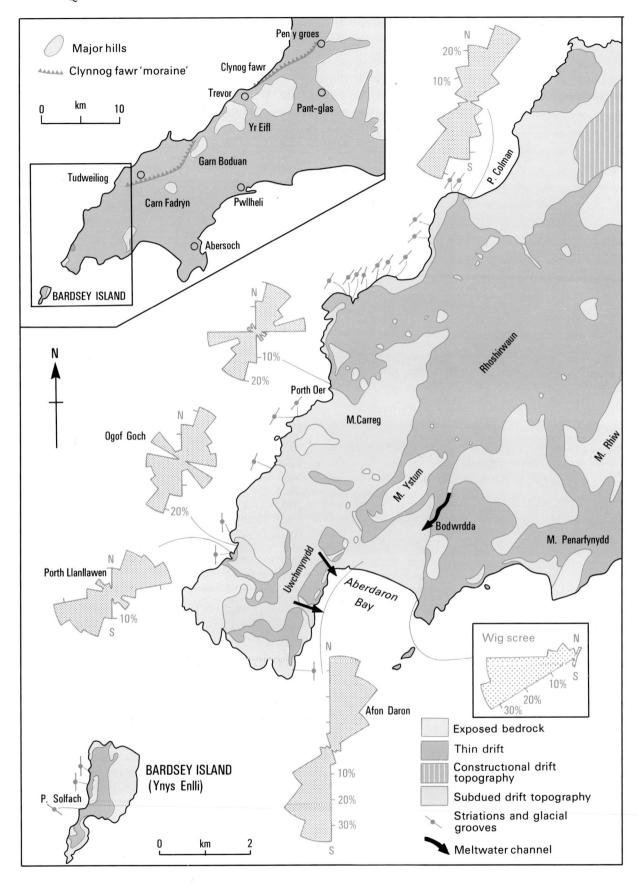

Figure 13 Evidence for ice-movement directions in western Llŷn.

of weakness. The Nant y Gadwen Valley [212 266] is similarly occupied by an underfit stream and may also have been formed subglacially. The small, completely drift-filled valleys which cut NE–SW across headlands at Porth Iago [168 317] and Porth Ysgaden [219 374] may have a similar origin.

Inland, to the north-east of Bodwrdda [189 272], the Afon Daron follows a fault-controlled valley which cuts through the thinly drift-covered bedrock ridge running north-south from Mynydd Ystum [186 285] to Penrhyn Mawr [190 262]. This ridge forms the eastern boundary of the thick drift around Aberdaron and its breaching probably indicates meltwater erosion beneath a glacier flowing from the north-east.

QUATERNARY DEPOSITS

Interglacial raised beach

The only deposits in western Llŷn which have been assigned to a previous interglacial are the cemented sands and gravels which occur in the drift section at the northern end of Porth Oer (Whistling Sands) [1668 3011]. The section was first noticed by Jehu (1909, p28) and was more fully described by Synge (1964), who recognised very hard, current-bedded, 'ferricreted' sands overlying well-rounded beach gravel of local origin, resting upon the raised marine platform. Whittow and Ball (1970, p.23), in their regional summary, accepted Synge's interpretation, describing a beach deposit 'containing a few erratic pebbles, perhaps from an otherwise unrecorded earlier glaciation'.

The recognition of interglacial raised beaches underlying drift deposits is of considerable importance in Quaternary geology. They indicate past sea levels and provide stratigraphical markers which facilitate correlation of the overlying deposits. The 'raised beach' deposits of Porth Oer, together with those of Red Wharf Bay on Anglesey, are the only examples in North Wales. They have been correlated with the 'Patella' beaches of Gower, which are of last (Ipswichian) interglacial age (Oxygen Isotope Stage 5e, about 120 000 BP; Bowen and Sykes, 1988) and which provided the basis for correlation of the drift deposits of North and South Wales (Bowen, 1973, 1977).

However, close examination of the present outcrop of the cemented sands and gravels at Porth Oer (Plate 23) reveals a more complex stratigraphy than that described by Synge (1964) and suggests that their interpretation as interglacial deposits may be incorrect.

The Porth Oer sequence appears to have been assigned interglacial status for three reasons: the sands and gravels lie directly upon the raised marine platform, they are correlated with sands which underlie the glacial sequence to the east and, like the Gower raised beaches, they are cemented by calcite. Re-examination of this section, and of the wide variety of drift exposures around the coast of western Llŷn, suggests that these criteria do not necessarily indicate a preglacial littoral origin.

The cemented sand and gravel does not rest directly upon the raised marine platform, but is underlain by similar but uncemented facies and by a diamict unit. Also, there appears to be no basis for correlating the cemented sands with the uncemented sands in the drift section to the east. Even where sands and gravels do rest directly on bedrock and are overlain by glacial deposits, it cannot be assumed that they predate glacial deposits which rest on bedrock elsewhere, as displayed in the small bay [2210 3754] between Porth Ysgaden and Porth Llydan.

Calcareous cementation is similarly not restricted to raised beach deposits. At Porth Oer, cement is derived from the calcareous Irish Sea drift deposits, and its mobilisation and deposition is the result of throughflow. Calcareous cementation is very common in drift sequences of western Llŷn and, although the most common cemented facies are sands and gravels, cemented diamicts also occur. A particularly well-cemented example can be seen south of Porth y Pistyll [160 246]. Some fine examples of cemented gravels occur in Aberdaron village [1734 2644].

Thus, there is no firm evidence to support the interpretation of the cemented deposits of Porth Oer as an interglacial raised beach. It is more reasonable to interpret them as Late Devensian or younger in age.

Details

The Porth Oer 'raised beach'

The cemented deposits lie at the north end of Porth Oer [1668 3011], on a small knoll separated from the main cliff by a track leading to Methlem Farm. They occur above a small fault-controlled cave in pillow lavas within the Gwna Mélange. Directly above the cave lies 40 cm of matrix-supported diamict, comprising locally derived angular to subrounded clasts up to 50 cm in diameter, together with a few far-travelled, smaller, rounded clasts in a muddy matrix. This is overlain by 10 cm of fine- to medium-grained, poorly sorted sand with larger clasts of the underlying deposits protruding into it. Above the sands lies 130cm of matrix-supported, crudely bedded gravel comprising well-rounded stones of mixed lithologies, up to 4 cm in diameter; the matrix is a muddy grit. In the lower portion, these gravels are mostly uncemented, though there are some small lenses with a degree of calcareous cementation. Higher in the section, the gravels become increasingly well cemented, forming a distinct overhang where they join the overlying medium- to coarse-grained, cross-bedded and extremely well-cemented sands (10cm). A further 10cm of sand overlies this and, although uncemented, is identical to and clearly a continuation of the underlying bed, because cross-stratification is continuous between cemented and uncemented sand. Above the sand, a thin (5cm) band of grey mud underlies the soil.

To the east, the gravel appears to thin and the degree of cementation in both sand and gravel decreases very rapidly. Two metres away, the facies are identical but completely uncemented. To the west, the gravel appears to die out rapidly and cemented sands lie directly upon bedrock.

Bay between Porth Ysgaden and Porth Llydan [221 376]

At the eastern end of the bay, directly overlying the raised marine platform, 40 cm of bedded, medium-grained sand is overlain by 90 cm of bedded gravel (clasts 1 to 3 cm). This is overlain by 10 cm of coarse sand, followed by over 2 m of stony mud with a few lenses of sandy gravel. However, just one metre to the west the bedrock dips beneath the beach, exposing a se-

Plate 23 Calcareously cemented sand and gravel underlain by uncemented sand, gravel and diamict at Porth Oer [1668 3011]. This sequence has previously been interpreted as an interglacial raised beach, but is more likely to be Late Devensian or younger in age. (A15044).

quence of deposits which underlie the lateral continuation of the sand and gravel.

Immediately beneath the sand is 20 cm of sandy gravel, which is in turn underlain by 30 cm of coarsening-upwards, brown sandy mud to muddy sand. At the base of the section lies an unknown thickness of blue-grey stony mud (typical 'Irish Sea till').

Pre-Late Devensian scree and head

Some deposits in western Llŷn appear to predate the last glacial advance. These can be divided into calcareously cemented, dolerite head deposits and uncemented screes.

The cemented head deposits are restricted to localities where dolerite sills crop out and are overlain directly by drift. The best and most accessible examples are found at Trwyn Bychestyn [166 262]. On the western side of this small headland, the dolerite sill is overlain by 0.5 m of uncemented head, comprising angular to subrounded dolerite clasts with a silty matrix. This is overlain by 1.5 m of identical but very well-cemented head which forms a distinct overhang and a small cave. The cemented bed displays calcitic flowstone depositional features. A similar sequence occurs on the east side of Trwyn Bychestyn, and less accessible examples occur on Maen Gwenonwy [2008 2598] and west of Porth Simdde [1666 2629].

Since these deposits comprise only immediately local material and are overlain by glacigenic sediments, they are interpreted as head; they probably reflect the periglacial conditions which preceded the last glacial advance.

A good example of an uncemented scree overlain by Quaternary glacigenic deposits occurs at Wig [1859 2582], at the easternmost end of Aberdaron Bay. Up to 10 m of stratified breccia is banked against a buried cliff. The deposit dips at 18° to 20° and generally fines upwards, although large blocks occur throughout. Towards the base, the deposit is clearly clast-dominated, with a laminated silt matrix which may have been washed in postdepositionally. The proportion of matrix increases up the sequence, and it also becomes more sandy. The top layers of the breccia merge into a diamict unit which is similar but which has a much greater proportion of matrix. This deposit forms a wedge thickening and dipping at 18° to the west, where it extends beneath the present beach. It is overlain conformably by a sequence of glacigenic sediments which are described later.

This deposit is interpreted as a scree which accumulated in situ, banked against a cliff, prior to the last glacial advance. The general fining and increasing proportion of matrix upwards may reflect increasing solifluction. Since the beds dip beneath the present beach, they must have been deposited when sea level was lower than at present. The sequence must have been protected from erosion by ice moving from the north-east because of its position on the west side of the prominent ridge which forms the eastern boundary of the Aberdaron embayment. Had the ice moved from the north-west, these deposits would have been removed.

Another fine example of a stratified scree deposit preserved in a sheltered position occurs at Porth Ysgo [2083 2653], where the deposits dip towards the south and were protected from erosion by their position in the lee of a fossil cliff. They also dip beneath the present beach and are overlain by glacigenic deposits. A spectacular 10 m section has been cut through these deposits by the waterfall of the Ysgo stream (Plate 24).

Late Devensian glacial deposits

Exposures of the Late Devensian glacial deposits of the district display a wide variety of lithological associations. Facies range from mud with few stones to silty and more sandy diamicts, which grade into gravel and sand. Rapid vertical and horizontal facies changes are typical, but are not reflected in the surface topography. It is not possible, therefore, to isolate and map individual lithofacies on the basis of their surface expression.

Sampling by auger or shallow drilling is of little value because of rapid facies change. Thus, all of the glacial deposits have been assigned to a single mappable unit, namely 'glacial deposits undifferentiated'. However, the glacial deposits, which have been mapped only where they are

Plate 24 Stratified local scree deposits incised by the waterfall of the Ysgo stream at Porth Ysgo [208 265]. The deposits dip below the present beach and are overlain by glacigenic sediments (visible top centre). The scree probably reflects periglacial conditions preceding the last ice advance. It was protected from glacial erosion because of its position in the lee of a fossil cliff. (A15042).

more than one metre thick, are classified on the basis of landform assemblages, which provide some indication of thickness and which aid reconstruction of environments of deposition. There are four types, the boundaries between which are often gradational and difficult to define.

A. *Thin drift where bedrock is the dominant control of topography* This category includes much of the centre of the district and the coastal strip of Ordovician rocks on either side of Aberdaron Bay. Although bedrock exposure is very limited over much of this area, particularly between Mynydd Ystum and Mynydd Cefnamwlch, the topography is characterised by long slope elements and relatively even surfaces, which contrast with the irregular undulating topography of the thicker drift.

B. *Areas of irregular or undulating topography where the bedrock has little influence on landform* These occur in three large tracts. The most extensive forms a broad swathe running south-westwards from Rhydlios to Uwchmynydd and grades into the thicker drift exposed in Porth Oer and Aberdaron Bay. A smaller area of thick drift lies between Mynydd Ystum and Mynydd Rhiw; it is bounded to the south by a ridge of Ordovician rocks mantled by thin drift. The third area is in the north, between Mynydd Cefnamwlch and Traeth Penllech. A terrace of thick drift also extends along much of the coast between Traeth Penllech and Porth Oer, along the west side of Aberdaron Bay, and between Trwyn y Penrhyn and Trwyn Talfarach.

Over much of the area the thick drift displays numerous small, enclosed depressions varying from 20 to only 3 m in diameter. Some form small pools, but most are dry. The best examples lie on the thick drift north-west of Aberdaron and east of Pencaerau [200 272]. They are not restricted entirely to the thick drift; Pwll Diwaelod [1642 2626], for example, lies on a bedrock ridge covered by relatively thin drift. Some of the smaller, shallow depressions may have been dug; at Ty Fwg [1682 2843], for example, the house is built of mud and straw, the mud having been obtained from the adjacent field, leaving a shallow, irregular pond which has since been filled in. A round, deeper pool in an adjacent field [1696 2849] is probably natural. One boggy enclosed hollow within a group, west of Pont Afon Saint [1634 2667], was augered and found to contain more than 2 m of peat and organic mud; others are dry and contain no peat. Most of the enclosed depressions can be interpreted as kettle holes produced by the melting of blocks of buried ice during deglaciation.

The unusually fine preservation of these features is probably due to the retention of the medieval field pattern over much of the area. Where field boundaries have been removed and more intensive farming methods employed, particularly in the northern and eastern parts of the district, the enclosed depressions can be seen in various stages of destruction. In Uwchmynydd, east of Porth Llanllawen [148 264], for example, three depressions which formed small pools have been drained, leaving only small crescentic hollows. To the east of Ty'n-lôn-fach [217 276], the field boundaries have been removed and the kettled surface has been drained and ploughed, resulting in an undulating topography which will eventually be destroyed by ploughing.

C. *Very thick drift filling preglacial embayments and, at the coast, extending beneath present beach level* Large areas of very thick drift, extending beneath present beach level, are restricted to Porth Oer and Aberdaron Bay. The drift surface in these areas is gently undulating and kettle holes are absent. The inland boundary of this category is gradational and difficult to define.

D. *Constructional topography: where drift is arranged as ridges and mounds* This landform assemblage is restricted to a very small area in the district, just to the north of Mynydd Cefnamwlch, where it is shown on the 1:50 000 map as **Morainic deposits** (see Young et al., in press). Ridges and mounds of drift continue to the east, however, to form the 'Clynnog Fawr moraine' which Synge (1964) suggested might mark the southern limit of Late Devensian Irish Sea ice in Wales. It is now widely accepted that Late Devensian ice reached much farther south (Bowen et al., 1986) and the 'moraine' probably represents a stillstand of active marginal ice banked against Mynydd Cefnamwlch and against Carn Fadryn, Garn Boduan and the other hills of the north Llŷn coast. It need not represent a re-advance.

The coast of western Llŷn provides many kilometres of drift exposure, but much of it is partly obscured by slumping. The thickest and most continuous sequence is exposed in Aberdaron Bay where it is being actively eroded.

Details

Aberdaron

The largest and most complex drift section in the area is that exposed in Aberdaron Bay, from Porth Simdde [1668 2632] to Wig [1858 2585] (Plate 25). Bedrock does not crop out anywhere along the beach and the thickness of drift is unknown. The gently undulating drift surface behind the coastal section is incised by the Cyll-y-felin and Daron streams. Remnants of the surface, which must have extended out into the bay, remain on its western side as a drift 'terrace' resting on bedrock. There are no kettle holes on the surface of the very thick drift between the Aberdaron cliff section and the outcrops of bedrock or thinner drift which surround it, although they are common on the drift surfaces to the north-west and beyond the thinner drift to the east.

Coastal protection works have obscured much of the western part of Aberdaron cliff and only the eastern section is described here (Figures 14 and 15). Distances are quoted in metres east of the boundary of the revetments [1756 2632]. Beyond 630 m the section is obscured by a large rotational slip and only the eastern 50 m, at Wig, are well exposed.

The main section, west of the large rotational slip, can be interpreted in terms of two main components; a lower diamict with abundant deformed silt bands and a sequence of upper diamicts with associated sands, gravels and laminated muds. The junction between them is irregular, reaching 8 m above the beach between 280 and 300 m, and between 460 and 490 m, but falls to within 2 m of the beach at 380 m and 550 m. West of 245 m, the contact is not exposed above the beach.

LOWER DIAMICT   The dominant facies of the lower diamict is a mud with scattered, often striated clasts and broken shell fragments. Contorted layers and lenses of very well-sorted white silt/fine sand occur throughout, ranging from a few centimetres to a metre thick and up to 4 m long. Some of the larger silt/sand bodies include ripple cross-stratification. Most of the

Plate 25 Aberdaron Bay. Glacial drift deposits in cliff section at Aberdaron Bay. The cliff maintains a steep angle and is being eroded mainly by falls and gullying. To the east of Aberdaron village, cliff stabilisation has been undertaken to protect the church. (A15036).

lenses have been folded and faulted. Between 460 and 490 m, sand beds towards the top of the lower diamict are less deformed and may have been redeposited.

Channel gravels up to 4 m wide and 2 m thick occur within the topmost layers of the lower diamict between 290 and 330 m. Gravels also occur within the lower diamict between 440 and 450 m, where they dip to the west and probably represent a channel about 10 m wide and 4 m deep, which has been post-depositionally tilted.

Where the upper diamict rests directly upon the lower diamict, the boundary is locally unclear. Over much of its length, however, they are separated by sorted sediments (Plate 26). Between 390 and 430 m, contorted silt bands are truncated and overlain by flat-based convex lenses of gravel. Only at one point (412 m) has the gravel eroded down into the silt. At 406 m the flat base of a gravel cuts across a silt body and the diamict on either side of it without regard to the marked difference in facies. Farther west, between 280 and 330 m, the junction of the lower and upper diamicts is marked by a sequence of eastward-dipping, well-sorted, rippled sands and cross-bedded gravels.

UPPER DIAMICT SEQUENCE   The upper diamicts are only crudely stratified, and contorted layers and lenses of fine sand/silt are absent. The dominant facies is a mud with scattered clasts and shell fragments. In small sections it appears massive but, from a distance, lines of cobbles and layers with a higher proportion of clasts define a crude stratification. East of 280 m, the stratification is approximately horizontal, but between 230 and 280 m it dips to the west. Variably sorted sands and gravels are common within the upper diamict sequence. To the east of 230 m, they are generally thin and discontinuous. Well-sorted sands and clast-supported gravels grade laterally into clast-dominated diamicts, which grade in turn into muddy diamict. West of 230 m, sands and gravels are better sorted and more continuous.

At 230 m, thin coarse sand overlying eroded diamict thickens westwards and grades into well-sorted sand and gravel, which in turn grades gradually into diamict. The sorted sediments are overlain by diamict which thins westwards to 194 m, where it wedges out and is overlain by horizontal sand and gravel. Farther west, between 30 and 120 m, the section, though badly slumped, is dominated by sorted sediments with a broad channel form, which are overlain by diamict.

RECONSTRUCTION   The Aberdaron section is interpreted in terms of stagnation of ice and the redistribution of sediments into an embayment now represented by the area of very thick drift and the present bay. The lower diamict is considered to be a melt-out till produced by slow in-situ melting of glacier ice

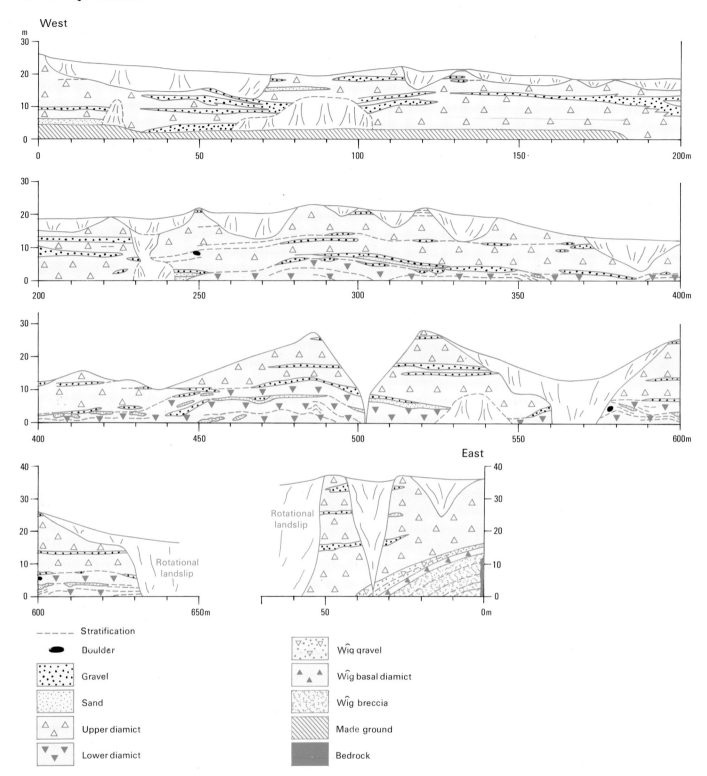

Figure 14 Quaternary deposits exposed in cliffs east of Aberdaron.

under a thick cover of sediments. Parts of the deformation within this unit probably occurred within the glacier prior to stagnation. Further deformation occurred during melting. The channel gravels and possibly redeposited silts near the top of the lower diamict may represent the deposits of englacial streams flowing near the surface of the decaying ice. The sorted sediments, which in many places form the junction of the upper and lower diamicts, probably formed as braided channels on the glacier surface.

The upper diamict sequence can be interpreted as an accumulation of redeposited diamict or flow tills derived from stagnant ice surrounding the embayment. The sorted sands and gravels were deposited by ephemeral braided streams. In many places diamict grades gradually into gravel, reflecting the dis-

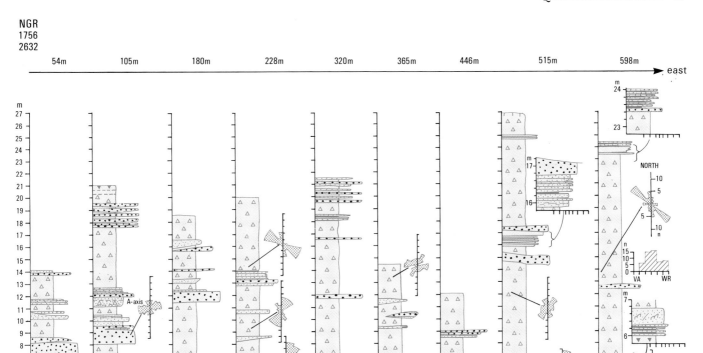

Figure 15 Selected graphic logs of sections exposed in cliffs east of Aberdaron (key as in Figure 14).

aggregation of flow tills as they entered channels. Clast-rich layers within the diamicts represent washing out of fines where insufficient water or energy was available to form channels and associated gravels.

The dip of the crude stratification and of sorted layers within the upper diamict sequence indicate that the underlying ice was continuing to melt during deposition (Figure 16). Between 300 and 440 m, the surface of the lower diamict forms a broad depression which is filled by upper diamict and capped by a horizontal sorted layer. Where these sorted sediments continue to the east, however, they dip down parallel to the surface of the lower diamict, suggesting that melting of the underlying ice occurred later at this point. Similarly, between 200 and 280 m, diamict fills a broad depression and is partly overlain by a horizontal gravel. The thicker sand and gravel deposits between 30 and 120m represent a more continuous channel system. Diamict layers wedging into the gravel reflect alternating dominance of flow-till and stream activity. Following abandonment, channels were filled by flow-tills, now represented by diamict with discontinuous sand and gravel layers.

Afon Daron

A smaller exposure in the thick drift of Aberdaron Bay occurs in the valley of the Afon Daron, 300 m south-west of Bodwrdda [186 271]. A meander of the river has eroded an 8 m section of the southern bank, revealing a complex sequence of laminated silts, diamicts and sands. At the west end of the section, which does not reach bedrock, the base comprises a blue-grey mud with scattered rounded clasts. Overlying this basal diamict are 4 m of interbedded, laminated silts and rippled sands, separated by an erosional boundary from 40 cm of poorly sorted gravel with lenses of diamict and sand. This is overlain by one metre of horizontal laminated silt, 65 cm of sand and an upper, more sandy diamict (80 cm).

The whole sequence here probably lies within the upper diamict sequence of the Aberdaron Bay section. The interbedded laminated sands and silts were probably deposited in a small lake. The climbing ripples in this sequence indicate flow towards a range of directions between north-west and north-north-east, that is towards, rather than away from the glacier, suggesting that the source of input was stagnant ice rather than an active ice front.

At the west side of the section, the beds dip to the east near the base, but the angle of dip declines up the sequence, and the beds become horizontal near the top. The ripple sets and silt drapes, however, must originally have been deposited subhorizontally, so that the sequence must have suffered deformation whilst the beds were being laid down (to account for the decline in dip). This deformation, and indeed the presence of an enclosed hollow in which a small lake might form, can be explained by the melting of buried ice. Flame structures occur at the junction of the basal diamict and the lowest sands.

The sand units display a general coarsening upwards, from fine-grained sand at the base to medium-grained sand higher in the sequence, and their colour changes from grey-blue to dark brown. Within the silt layers, and less commonly in the sands,

Plate 26 The Aberdaron Bay section between 480 and 530 m east of the revetments. The lower and upper diamicts are separated by well-sorted sands which dip to the east.

there are isolated clasts up to 5 cm in diameter and small clusters and layers of gravel. These may represent debris released from lake ice rather than iceberg dropstones.

The gravels overlying the lake deposits contain 'clasts' of uncemented sand, which must have been incorporated whilst frozen, and two lenses of silty gravel, one of which displays a clear flow nose with concentric lamination. They probably represent flow tills which have entered a braided stream system.

The overlying sequence of mud followed by sand can be interpreted in terms of relatively minor changes in the energy of input to the site. The section is capped by a crudely stratified diamict comprising subangular to well-rounded clasts up to 20 cm in diameter in a silty sand matrix. It probably represents the redistribution of diamict as flow till or solifluction from the last remnants of decaying ice in the surrounding area.

Wig

At Wig, the easternmost end of the Aberdaron Bay section, beyond the large rotational slip, the lower diamict and associated deformed silt bands appear to be absent. The scree and solifluction deposits are overlain by a sequence of diamicts and sorted sediments. The first 'foreign' unit is 2 m of sandy mud with abundant striated clasts of mixed lithology. The surface of this diamict is eroded beneath 3.5 m of pebble to cobble gravel with a muddy sand matrix. Many of the clasts in this gravel retain striations. The remaining 37 m of the section comprise facies similar to the upper diamict sequence further west. The beds at the base of the sequence dip conformably with the underlying scree and thin to the east, but higher in the sequence the sorted layers are approximately horizontal.

The absence of the lower diamict sequence at Wig suggests that stagnant ice was not buried by a deep cover of sediment this close to the edge of the embayment. The lowest diamict may have been deposited subglacially, but the remainder of the sequence probably represents flow tills and stream deposits derived from the higher land to the east.

Porth Oer

At the northern end of Porth Oer, an exposure shows 8m of drift (Plate 27) overlying the weathered pillow lava discussed earlier. Bedrock, exposed only at the northernmost end of the section, is overlain by a thin layer of local clasts in a sand matrix, in turn overlain by 2 m of bedded medium- to coarse-grained sand, including a few isolated pebbles and cobbles up to 20 cm in diameter and some beds, up to 10 cm thick, of diamict with a silty sand matrix. The sand unit thins to the west and is only 10 cm thick above the bedrock; to the east, it thickens and the base becomes obscured. The sand is overlain, in the centre of the section, by a stratified sandy diamict comprising mainly local, angular clasts with a few well-rounded pebbles from elsewhere. This bed is eroded and overlain by 1 m of very stony diamict with more foreign clasts and a silty gravel matrix. The surface of both diamicts is eroded and overlain by 80 cm of fine- to medium-grained dense sand containing a few small stones. This is followed by 2 m of crudely stratified diamict, including clasts up to 50 cm in diameter in a muddy sand matrix. This diamict becomes more stratified and contains more clasts upwards, eventually grading into 1.2 m of gravel with abundant well-rounded pebbles. This is overlain by 60 cm of bedded sand which includes small pebbles, comminuted shell fragments and occasional thin gravel lenses. This bed thins to the north and the underlying gravel becomes overlain directly by one metre of grey mud, which includes small rounded pebbles, occasional small discontinuous layers of sand and some comminuted shell.

The sequence can be interpreted in terms of flow tills grading into braided stream gravels. To the east, the thick glacial deposits of Porth Oer extend to an unknown depth below beach level and are poorly exposed due to slumping. The domi-

QUATERNARY DEPOSITS 69

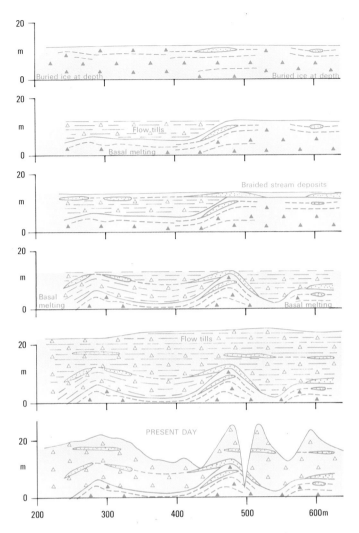

Figure 16 Stages of development of the Aberdaron drift sequence (key as Figure 14).

nant facies are similar to the upper diamict sequence of Aberdaron Bay.

Porth Ysgo to Porth Llawenan

Along the section from Port Ysgo [208 265] to Porth Llawenan [214 260], the sequence is difficult to interpret because of slumping. It is clear, however, that the scree deposits are overlain by glacigenic sediments, as at Wig. However, the deposits here are unique in the district, in that they include boulders up to 5 m in diameter, many of which have been eroded out of the drift and now lie on the beach. The large boulders are angular to subangular, and were probably derived locally and transported supra- or englacially rather than at the glacier sole.

Porth Ysgaden

In the extreme north of the area, at Porth Ysgaden [219 374], a small south-west-facing, rock-bounded bay reveals a 6 m drift section in a small buried valley. The base of the section does not reach bedrock but exposes a blue-grey mud with scattered clasts. Above this, 15 cm of coarse gravelly sand has eroded the lower diamict surface. This sand unit includes isolated clasts up to 5 cm in diameter. Above this lies 30 cm of coarsening-upwards fine- to medium-grained sand. This unit is distorted and crossed by high-angle normal faults, perhaps due to melting ice in the underlying beds or to loading from above. These sands contain lenses of blue-grey mud up to 50 cm thick, similar to the basal diamict and incorporating occasional isolated clasts up to 10 cm in diameter. Above this, 85 cm of bedded coarse sand and gravel similarly includes lenses of mud. Overlying the gravel are 20 cm of blue-grey mud which includes isolated clasts, up to 10 cm in diameter. The surface of this bed is eroded and overlain by 60 cm of sandy, bedded gravel with clasts 1 to 15 cm in diameter, many of which are clearly striated. This is overlain by another, thicker bed of grey mud (1.6 m) beneath a final stratified sandy mud with occasional clasts.

The sands and gravels overlying the basal diamict can be interpreted as braided channel deposits and the muds as flow tills.

Penrhyn Mawr and Porth Colmon

At a few north-facing coastal localities, a distinctive lithofacies is preserved, comprising clearly stratified muds with abundant pebbles, comminuted shell fragments and some small pieces of blackened wood. They probably represent marine deposits entrained at the glacier sole and redeposited subglacially on north-facing rocky coasts, the stratification resulting from shearing due to overriding. Good examples can be found on the north coasts of Penrhyn Mawr [322 170] and Penrhyn Colmon [193 343].

Interpretation of glacial sediments

All of the glacigenic deposits described here, together with the erosional and depositional landforms, are interpreted in terms of the advance and subsequent decay of a terrestrial ice sheet. The striations suggest that at some stage the Irish Sea glacier was not frozen to its bed over western Llŷn but was sliding across it. The amount of erosion achieved by subglacial abrasion was limited, however, since the coastal platform is striated but not destroyed, and scree deposits which must predate the last advance remain banked against south- and south-west-facing buried cliffs. Erosion by subglacial meltwater was considerably more effective, carving steep-sided valleys, although the extent to which these have developed over successive glaciations is unknown.

Although these features suggest warm-based conditions, with melting and sliding at the base, the drift geology of the area is not typical of the landform-sediment assemblage often associated with such an ice sheet. Lodgement tills are not widespread and there are no drumlins or other streamlined drift landforms. Rather, thick glacial deposits with an irregular and undulating topography characterise low areas, whilst gentle slopes support a thin drape of sediments, and steeper slopes and summits are largely denuded.

The drift landforms and most of the deposits are best explained in terms of the stagnation and in-situ decay of the ice sheet and the redistribution of sediments downslope as flow tills, which accumulated on the low land or were disaggregated and removed in braided stream systems. In Aberdaron Bay and perhaps Porth Oer, debris-rich glacier ice was buried by a thick cover of flow tills as it melted, producing an undulating contact between

Plate 27 An 8 m drift section exposed at the northern end of Porth Oer [1677 3011]. The pillow lavas at the base are weathered to a depth of 2 m. The overlying sequence of sands, gravels and sandy diamicts can be interpreted in terms of flow tills grading into a braided stream system. (A15046).

melt-out till (lower diamict) and flow tills (upper diamict sequence) beneath a gently rolling land surface (Figure 17). Elsewhere, the 'dead ice topography' produced by the melting of ice beneath a thinner cover of sediments is still evident in the form of enclosed depressions (kettle holes). The lowland drift surfaces, produced by the accumulation of sediments flowing from higher land, must have extended beyond the present coast. The remnants of this surface are still evident in the form of a coastal drift terrace.

Head deposits

The term 'head' is restricted here to locally derived material which has been transported and deposited in a periglacial environment. It is not used to include glacigenic deposits which have been postdepositionally transferred downslope. If such a criterion was accepted, it would include most of the glacial deposits.

The only major landform in south-western Llŷn which is associated with 'head' forms a 'footslope' along the western side of Mynydd Rhiw, in the south-east of the district. These deposits are well exposed in a small quarry at Baron Hill [2190 2847], where bedrock is overlain by 1.2 m of sandy, matrix-supported diamict containing only local clasts up to 15 cm across. Some of the clasts display in situ spheroidal weathering. In many areas along this 'footslope', particularly around the village of Rhiw [225 275], large boulders litter the surface, presumably rolling from the crags above or rafted by solifluction; in many places these have been cleared from fields, producing large piles of boulders.

QUATERNARY DEPOSITS 71

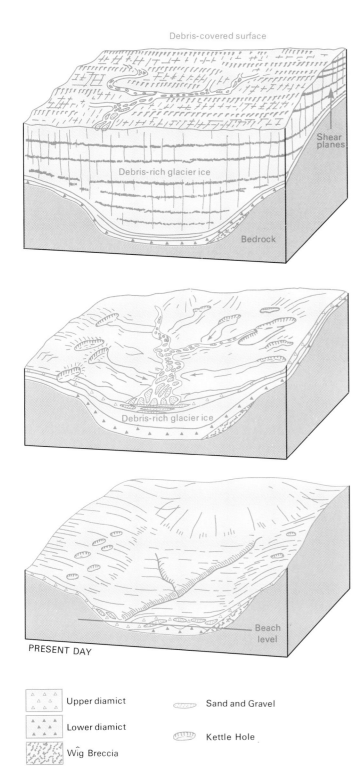

Figure 17 3-dimensional model of the development of the drift landforms around Aberdaron.

Thin, local slope deposits, which might be termed 'head', occur elsewhere in the district, but they are not of sufficient thickness to warrant inclusion on the map. Some of the better exposures can be seen in Uwchmynydd at Bod-isaf [151 268], on the west side of Porth Felen [143 250], in a small quarry on the northern side of Mynydd Ystum [179 279] and on the western side of Mynydd Penarfynydd [218 267].

Postglacial raised beach

At Porthorion [1564 2880], cemented raised beach deposits are preserved in a cave in the Gwna Mélange. The base of the beach lies at about 2 m above OD and the cemented deposits are about 1m thick. The beach gravels, with associated sand and shells, have partly blocked a horizontal fissure and incorporate local angular boulders up to 50 cm in diameter. The gravels are rich in mollusca; four whole specimens of *Patella vulgata* and two of *Gibula umbilicalis* were submitted for amino-acid analysis. The very low (<0.036) D-alloisoleucine/L-isoleucine ratios and the absence of free amino acids indicate that the shells do not predate the Holocene (D Q Bowen, personal communication).

Alluvium

Postglacial deposits of interbedded silts, sands and gravel occur along the three main streams in the district: the Afon Cyll-y-felin, from Aberdaron [173 264] to Ty Nant [173 283]; the Afon Daron, from Aberdaron to Bodwrdda [189 272] and beyond the gorge as far as Bodrydd [203 285]; and the Afon Fawr, on either side of Pont yr Afon Fawr [206 341].

The Afon Daron and the Afon Cyll-y-felin both display some evidence of incision into higher terraces, but these are fragmentary and have not been accurately levelled. The best examples occur along the Afon Cyll-y-felin to the east of Ty-isaf [171 278].

Landslips

Active and fossil landslips occur on many of the drift cliffs of western Llŷn. Only one, towards the eastern end of Aberdaron Bay [184 259], is sufficiently large to be included on the map. This is a rotational slide with a marked headwall. A smaller example occurs towards the northern end of Porth Oer.

Elsewhere along the coast, most landslips are much smaller and of a composite nature; they often began as rotational slides, with a distinct headwall, but degenerated into flows. This composite nature reflects the different mechanical properties of the lithofacies which comprise many of the drift sections, and the concentration of groundwater above relatively impermeable horizons. Various composite slides and flows occur along the coast west of Mynydd Carreg [158 293], and a particularly fine example of a mudflow can be seen at Porth Oer [167 301].

The only area where cliff erosion forms a serious threat is along the beach to the east of Aberdaron village, where the road from Aberdaron to Rhiw lies close to the cliff top. For most of this section the drift cliff maintains a high angle and appears to be retreating mainly by falls and gullying, with some small-scale rotation evidenced by small scarps behind the cliff top. Cliff stabilisation, drainage and buttressing have been undertaken to protect the church.

QUATERNARY HISTORY

Preglacial landscape evolution

The large-scale features of western Llŷn probably predate the Pleistocene glaciations, but their origin and the timescale involved in their evolution is the subject of debate. The classical interpretation of Whittow (1957) and Brown (1960) suggests marine erosion of a Tertiary peneplain. There are, however, several problems with accepting this model of marine planation and the associated pulsed 'eustatic' uplift concept of landscape development. Modern process studies have questioned the ability of marine erosion to cut a broad plain (King, 1963; Ollier, 1981) and uniform erosion of landscapes under humid temperate conditions (as envisaged in the Davisian cycle of erosion) is also in doubt (Twidale, 1976; Brunsden, 1980; Bradshaw, 1982). More seriously, the theory requires long periods of structural stability over wide areas, punctuated by short-lived tectonic events. Geophysical surveys in the Irish Sea Basin suggest active block-faulting during Cenozoic times, so long-term stability over wide areas is unlikely to have occurred.

More recently, Battiau-Queney (1980, 1984, 1987) has argued that etchplanation was the major process involved in shaping the preglacial landscape of Wales. Etchplanation involves deep weathering under humid tropical or subtropical conditions, followed by removal of the saprolite (weathered products) to reveal the weathering front as an 'etchplain'. The concept was first used to explain the formation of surfaces in the humid tropics (Wayland, 1934; Büdel 1957; Thomas, 1974), but has also been applied to surfaces in Europe (Büdel, 1979; Bremer, 1980), in Scandinavia (Lindmar-Bergström, 1982), in West Cornwall (Walsh et al., 1987) and in north-east Scotland (Hall, 1985, 1986, 1987). Etchplanation surfaces need not be related to a stable base level and are unlikely to be flat. Differential resistance of bedrock to weathering, either due to lithology or structure, and differential removal of saprolite, can lead to the formation of inselbergs (isolated hills), closed hollows and even broad valleys. The concept has been used to explain basin and ridge topography in the humid tropics (Thorp, 1967; Kroonenberg and Melitz, 1983) and even the topography of Dartmoor (Waters, 1957).

Battiau-Queney (1984), in common with other French workers (Pinchemel, 1969; Beaujeau-Garnier, 1975), considers that peneplains form on individual massifs in syn- or immediately postorogenic times. She suggests that an extensive planation surface was formed in North and Mid Wales during the Devonian. Anglesey, and probably western Llŷn, must have emerged by Triassic times to contribute to fine sediments recovered in the Mochras Borehole near Harlech (Harrison, 1971). The area then underwent a long period of subaerial conditions involving deep weathering and the etching of the major features of the present landscape. The break-up of the post-Caledonian surface to produce the present difference of relief between Anglesey and Snowdonia is placed in the late Tertiary, since it must postdate Eocene dykes which presently crop out at a low level in Anglesey and Arvon, but at 750 to 780 m in Snowdonia (Greenly, 1938). Battiau-Queney (1984) explains the differential uplift by 'warping' along a major hinge zone associated with the Menai Strait Fault System. Gibbons (1987) suggests that this fault system converges to form the Llŷn Shear Zone, i.e. the eastern boundary of the Gwna Mélange. The generally low relief of western Llŷn, compared to Snowdonia, may reflect close proximity to the hinge zone and perhaps a decline in degree of deformation south-west of the main centre.

Thus, it seems that the major bedrock landforms of the Aberdaron area, the isolated hills, the intervening broad valleys and the plains, may have been formed over a much longer timescale than has traditionally been assumed. The hills may have emerged as inselbergs, preserved because of their greater resistance to chemical weathering, whilst the weaker areas, more easily attacked because of less resistant lithologies or a greater density of discontinuities (e.g. joints and faults), were etched to a greater depth. However, during the Pleistocene, the area was inundated, probably on several occasions, by ice sheets moving down the Irish Sea Basin, and most of the direct evidence of previous landforming processes has been eroded or covered by drift. Moreover, the differences in relative resistance which are likely to lead to differential relief under tropical weathering are, in many cases, likely to provide similar differential resistance to glacial and fluvioglacial erosion. One small outcrop of weathered pillow lavas is insufficient to confirm the concept of widespread etchplanation. A greater understanding of the long-term, preglacial evolution of the landscape of the Aberdaron area must await a re-emphasis by British workers on such problems on a regional scale, and on the influence of passive-margin tectonics on geomorphology (Thomas and Summerfield, 1987).

Raised rock platform

The formation of rock platforms, such as that around the coast of western Llŷn, has traditionally been assigned to marine erosion during periods of relatively high interglacial sea level. The suggestion that platforms may have been formed during successive periods of interglacial marine erosion has been criticised on the basis that Pleistocene sea levels have fluctuated rapidly (Dawson et al., 1987). Recent work suggests, however, that during interglacials corresponding to Oxygen Isotope Stages 5e, 9 and 11, marine limits were probably very close to their present level (Shackleton, 1987).

An alternative explanation is that the platforms were formed not by wave action in relatively warm interglacials, but by periglacial shore processes. The idea was first put forward by Fairbridge (1977) who suggested that early and mid-Pleistocene cold-phase sea levels were relatively high. Sissons (1981) and Sissons and Dawson (1981) argued for active periglacial erosion around the coast of Scotland and suggested that platform remnants farther south might have a similar origin, implying limited ice extent during some cold phases. A major obstacle to the acceptance of this theory has been a lack of evidence as to the efficiency of periglacial processes on hard rock shores. Recent work in western Norway, however, has demonstrated rapid formation of rock platforms on a lake shoreline under periglacial conditions (Matthews et al., 1986) and has been used as a modern analogue for the formation of both lake and marine platforms under periglacial conditions during the Pleistocene (Dawson et al., 1987).

Quaternary glaciations

As noted above, the glacial deposits of western Llŷn are considered in this account to record the advance and decay of a single ice sheet, which moved down the Irish Sea Basin. No direct evidence has been obtained within the area of the timing of the onset or decline of glacial conditions. Detailed oxygen isotope studies of deep sea cores, however, record numerous glacial/interglacial cycles over the last 2.4 million years (Shackleton and Opdyke, 1973, 1976; Shackleton et al., 1984). On land the record is less complete, but around the shores of the Irish Sea there is evidence for three advances.

The earliest, which reached Fremington on the north coast of Devon (Bowen, 1969) and deposited northern erratics in South Wales (Strahan and Cantrill, 1904), has been correlated with the Anglian glaciation of England and the Munsterian of Ireland (Bowen et al., 1986). It has been assigned to Oxygen Isotope Stage 12 (428 to 480 ka), which correlates with the Elster glaciation of Europe.

Evidence for the intermediate advance is limited to Paviland in Gower, where a recently identified moraine ridge appears to predate the raised beaches of Minchin Hole, which are ascribed to Oxygen Isotope Stage 7 (Bowen et al., 1986). The Paviland glaciation has been correlated with the Kirkhill till of Scotland (Connel and Hall, 1984) and the post-Hoxnian but pre-Ipswichian deposits of England (The term 'Wolstonian' is no longer regarded as appropriate since the deposits at Wolston are probably Anglian in age; Rose, 1987). It has been assigned to Oxygen Isotope Stage 8 (252 to 302 ka), which is the Drenthe (Saale) of Europe (Bowen et al., 1986).

The most recent glaciation of the Irish Sea Basin occurred during the Late Devensian (Oxygen Isotope Stage 2; 10 to 26 ka) and is known as the Dimlington advance in Britain (Rose, 1985) and the Glenavy advance in Ireland (McCabe, 1987a). Although some of the erosional landforms of western Llŷn may have developed during successive advances, all of the glacigenic deposits of the area can be assigned to the last, Late Devensian (Irish Midlandian) cold stage. The head and scree now covered by glacial sediments and in places extending below present sea level may represent periglacial conditions, with relatively low sea levels, prior to the last ice advance.

The extent of the Late Devensian ice in the Irish Sea Basin has been the subject of considerable debate. Mitchell (1960, 1972) and Synge (1964) placed the limit along the north coast of Llŷn, leaving the Aberdaron area and the southern Irish Sea unglaciated. Their methodology involved mapping moraine ridges, counting the number of 'till' units in coastal sections and assessing the degree of weathering of tills and the apparent 'freshness' of landforms. This methodology was criticised by Bowen (1973) who presented a simpler explanation of the glacial deposits surrounding the Irish Sea using interglacial raised beach deposits as stratigraphic markers.

Bowen (1973) correlated the raised beaches of Gower with similar deposits at Poppit near Cardigan, Porth Oer in the Aberdaron area and Red Wharf Bay on Anglesey. Since all of the glacial sequences north of Gower lie stratigraphically above these deposits, they were all assigned to a single 'post-raised beach' glaciation which he proposes to be Late Devensian. The southern limit of Late Devensian ice was placed at West Angle Bay near Milford Haven (John, 1970). Welsh ice reached as far south as Gower and Cardiff. This interpretation has subsequently been supported by amino acid geochronology, which places most of the Gower beaches in the Ipswichian (Oxygen Isotope Stage 5e) and has been used to recognise shells from Oxygen isotope Stage 3 in glacial drift from south-west Dyfed, north Gower and the Isle of Man (Bowen et al., 1986; Bowen and Sykes, 1988).

The timing of the Late Devensian Dimlington advance is well constrained at and near the type site on the East Yorkshire coast (Rose, 1985). Here glacial diamicts directly overlie silts which contain moss and coleoptera with glacial affinities and which have yielded radiocarbon dates of $18\,500 \pm 400$ BP and $18\,240 \pm 250$ BP (Penny et al., 1969). Lake deposits which overlie glacial diamicts at two sites have yielded radiocarbon dates of $16\,713 \pm 340$ BP (Jones, 1977; Keen et al., 1984) and $13\,045 \pm 270$ BP (Becket, 1977). The maximum age is supported by a thermoluminescence date of $17.5 \pm 1.6 \times 10^3$ years from a solifluction deposit underlying diamict of the Dimlington advance at Epplewoth in Yorkshire (Wintle and Catt, 1985).

The timing of the Dimlington advance in the Irish Sea Basin is not so well constrained and need not be synchronous with the east coast. Five radiocarbon dates in the range 18 400 to 18 900 BP obtained from the base of kettle holes overlying Dimlington advance deposits in the Isle of Man were initially used as evidence that the Irish Sea ice had retreated substantially before the ice on the east coast had reached its maximum extent (Mitchell, 1972; Thomas, 1976, 1977). However, the moss (*Drepanocladus revolvens*) which was dated is capable of subaquatic photosynthesis, so these anomalously old dates may be due to a hard-water error (Lowe and Walker, 1984). This is supported by a date of 18 000 (+1400–1200) obtained from a mammoth bone in a cave in the Vale of Clwyd which is sealed by Dimlington advance till (Rowlands, 1971). A radiocarbon date of $14\,468 \pm 300$ BP has been obtained from organic deposits overlying Dimlington advance deposits at Glanllynau, in eastern Llŷn (Coope and Brophy, 1972), although here the ice came from the mountains of North Wales rather than down the Irish Sea. The evidence suggests, therefore, that Late Devensian ice probably covered Llŷn soon after 18 000 BP and that the area was ice free by about 14 500 BP.

There is another piece of evidence which may further constrain the timing of deglaciation of the Irish Sea Basin. Radiocarbon dates of $16\,940 \pm 120$ BP and $17\,300 \pm 100$ BP have been obtained from molluscs considered to be in situ in glaciomarine deposits at 40 m above OD in north County Mayo, western Ireland. They have been compared with deposits at a similar elevation on the east coast of Ireland and used, together with sedimentological evidence, to suggest relatively early deglaciation triggered by an isostatically raised sea level (McCabe, 1986a, 1986b, 1987a, 1987b; Bowen and Sykes, 1988; Eyles and McCabe, 1989a, 1989b, 1989c).

Detailed reconstruction of the environments of deposition of complex drift sequences is a relatively recent field of study. Such sequences have traditionally been interpreted in terms of 'tills', representing glacial episodes, separated by sands and gravels, representing warmer, ice-free conditions. Although some early workers perceived more complex depositional systems (e.g. Goodchild, 1875; Lamplugh, 1903), a sound basis for glacial sedimentology was only provided by modern process studies. The environment of deposition of the drift deposits surrounding the Irish Sea is currently the subject of considerable debate. The discussion centres around whether the complex drift sequences represent terrestrial conditions involving lodgement, supraglacial meltout and proglacial fluvial and lacustrine deposition (Allen, 1982; Boulton, 1977; Dackombe and Thomas, 1985; Thomas, 1977, 1984, 1985; Thomas and Kerr, 1987; Thomas and Summers, 1982, 1983, 1984), or sedimentation in a glaciomarine environment, with high relative sea levels causing rapid retreat of the ice front (Colhoun and McCabe, 1973; Eyles and Eyles, 1984; Eyles et al., 1985; Eyles and McCabe, 1989a, 1989b, 1989c; McCabe, 1986a, 1986b, 1987; McCabe et al., 1984, 1986, 1987; McCabe and Eyles, 1988; McCabe and Hirons, 1986).

The glacial landforms and deposits of western Llŷn have here been interpreted in terms of the stagnation and in-situ decay of a terrestrial ice sheet carrying sediment derived from the floor of the Irish Sea. Although in accord with the sedimentological and geomorphological evidence, this model raises two important questions: how did debris become entrained in the ice and why did the glacier stagnate and melt in situ rather than retreat by melting at the snout?

Thermal conditions can vary in four zones (A–D) beneath an idealised ice sheet (Figure 18; Boulton, 1972). The clear evidence of glacial abrasion in the form of grooves and striations, and the presence of steep-sided gorges interpreted as subglacial meltwater channels, attest that this area of Llŷn must at some stage have been covered by ice in zone A. The incorporation of large amounts of material into the ice from a subglacial source (the floor of the Irish Sea) could have occurred in zone C, where water and sediment freeze on to the sole and shear up into the glacier. The erosional features and incorporation of the englacial sediments may, therefore, have been initiated under different basal thermal regimes and, therefore, at different times. Zone A conditions, with net basal melting are most likely to have occurred when the ice was at its thickest and most extensive. As the ice front retreated or the ice thinned, the

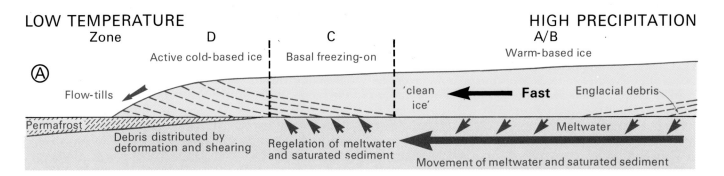

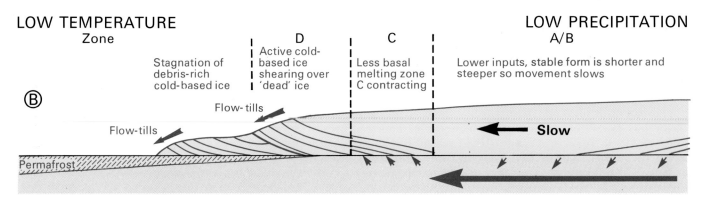

Figure 18 A model which facilitates the incorporation of subglacial sediments at a transition from warm-based to cold-based ice.

zone of net basal freezing (D) would migrate up-glacier. Subglacial sediments may have been incorporated by basal freezing during the passage of zone C, followed by stagnation of debris-rich cold-based ice.

Stagnation of the ice sheet and melting in situ may have been a local phenomenon. As the ice moving from the north-north-east thinned and the ice front approached the Aberdaron area, a critical point would have been reached where the ice no longer had sufficient surface gradient to carry it over the hills of the north Llŷn coast, causing stagnation of ice to the south and southwest. A prolonged stillstand of active marginal ice banked against these hills could explain the band of constructional drift topography known as the Clynnog Fawr moraine (Synge, 1964).

Alternatively, stagnation and melting in situ may have been more widespread. The classical interpretation of deglaciation is that higher air temperatures lead to greater ablation, so that the ice front retreats by frontal melting. The mass balance of a glacier depends, however, on precipitation as well as ablation, and the former may have been the dominant control on deglaciation. For an ice sheet resting on a deformable bed, when precipitation is increased, the stable long-profile is shallower, and so the glacier accelerates and thins (Boulton and Jones, 1979). The corollary of this is that if precipitation is reduced the glacier moves more slowly and the stable form is steeper. Boulton and Jones (1979) estimate that the response time of the Irish Sea glacier may have been in the region of 1000 years. An increase in precipitation may, therefore, have very gradually changed the form of the glacier. A rapid decline in precipitation, however, would lead to ice towards the margins being starved, resulting in a transition from warm-based to cold-based ice, and to stagnation.

This model of incorporation of marine sediments from the Irish Sea Basin accounts for many of the marine characteristics of the Irish Sea drift, including the dominance of muddy facies and the presence of comminuted marine shells. It assumes, however, that the deposits were released in a terrestrial environment. It has recently been suggested that much of the glacial drift around the margins of the British Isles may comprise glaciomarine sediments deposited by rapidly retreating ice fronts in an isostatically raised sea (Colhoun and McCabe, 1973; Eyles and Eyles, 1984, Eyles et al., 1985; Eyles and McCabe, 1989a, 1989b, 1989c; McCabe, 1986a, 1986b, 1987; McCabe et al., 1984, 1986, 1987; McCabe and Eyles, 1988; McCabe and Hirons, 1986). Under this model most of the deposits described here would be interpreted as submarine valley infill complexes, with marine muds, including iceberg drop stones, interbedded with gravels and diamicts, representing debris flows.

The glaciomarine hypothesis is difficult to reconcile with some of the landforms and sediments of western Llŷn. The section in the valley of the Afon Daron, for example, shows clear evidence of deposition in a small deforming basin, with sediment being supplied by currents moving towards the ice front. Similarly, the abundant evidence of buried ice, in the form of the undulating boundary of the upper and lower diamicts at Aberdaron and the numerous enclosed depressions on the surface of the thick drift, is difficult to explain in terms of a submerged landscape.

SEVEN

Economic geology

The district has been locally quarried on a small scale for sand and gravel, building stone, ornamental jasper, limestone, copper and baryte. All these minor activities were, however, eclipsed by the mines of Nant and Benallt, which together constituted the most important source of manganese ever mined in Great Britain.

BUILDING STONE, JASPER, AGRICULTURAL LIME

The lack of any large settlement in the Aberdaron area has avoided any requirement for major sources of building stone. Most farmhouses used whatever local stone was available, although the importation of foreign stones such as Carboniferous limestone from Anglesey also occurred. Massive igneous rocks, such as dolerite, and various members of the Sarn Complex were commonly quarried for building purposes. Examples of old quarries in these materials may be seen at Baron Hill [2191 2847] (dolerite), Pen Y Gopa [2223 3173] (granite) and Graig Fael [2171 3039] (gabbro). Jasper was quarried from clasts associated with basaltic lava in the Gwna Mélange. The best example of this is the relatively large disused quarry on the north side of Mynydd Carreg [1624 2921]. Limestone, used for agricultural purposes, was also won from clasts in the Gwna Mélange, as can be seen on the south-west flank of Mynydd Mawr [1374 2567].

COPPER, BARYTE

Unsuccessful attempts to find metallic ore have been made in the precipitous cliffs at the remote coast on the west side of Mynydd Anelog, where several horizontal shafts (adits) were cut along a vertical quartz vein, and also on the coast of Trwyn Talfarach [2170 2577]. In 1970–72 the layered intrusive rocks around Rhiw and Mynydd Penarfynydd were explored for copper and nickel mineralisation by Noranda-Kerr Ltd. Geochemical sampling revealed some anomalies for copper in soils, but in insufficient concentrations to justify further work. A more profitable mining exercise was achieved along the Afon Saint Fault, where an adit was driven along a quartz vein rich in baryte (barium sulphate). The ore was shipped from a jetty (no longer existing) at Porth Simdde [1670 2628], above which the adit is still traceable. The lode worked from the adit consisted of a pale intergrowth of baryte and quartz. Similar baryte-rich quartz veins, again associated with a fault, occur along the coast north of Ynys Piod [1653 2582]. This fault, which downthrows shattered dolerite of the Pen y Cil Sill against the laminated siltstones just above the Porth Meudwy Formation, is marked by a deep cave.

SAND AND GRAVEL

Layers and lenses of sand and gravel are a common component of the complex drift sequences exposed in the Aberdaron district. The coastal section at Aberdaron, for example, includes sands and gravels exhibiting a wide range of grain sizes and degrees of sorting. However, the heterogeneity of these complex drift sequences and the difficulty of locating sorted sediments at depth, where there are few inland exposures, means that they can only be exploited on a very small scale. For example, a thin, poorly sorted gravel has been extracted west of Garreg Fawr [1505 2477].

The greatest potential for exploitation of sand and gravel probably lies in the area of constructional drift topography in the north-east. This small area forms the western extremity of a band of constructional topography which extends along the north coast of Llŷn and which has been called the Clynnog Fawr moraine (Synge, 1964). Exposures elsewhere on this topography reveal thick sequences of often well-sorted sand and gravel.

MANGANESE

The disused manganese mines lie to the west of the village of Rhiw, one lying to the north of the Aberdaron-Rhiw road (Benallt Mine) and the other to the south (Nant Mine). Manganese deposits were worked on a relatively small scale between 1894 and 1904, with only a few hundreds of tons being produced per year. Mining activity increased dramatically between 1905 and 1907, with production up to 20 000 tons per year. For the next twenty years or so, most of the manganese ore came from the Nant Mine, until it was closed in 1925 (Woodland, 1956). Annual production during this time peaked at 9308 tons in 1918, at the end of the First World War. The total recorded output of manganese ore from these mines up to 1928 was 134 770 tons. This compares with 43 944 tons won from Cambrian manganese deposits in the Harlech area and 57 141 tons from Upper Palaeozoic strata in south-west England. These three areas constituted the only important sources of manganese known in Britain. Benallt Mine was reopened during the last war, from 1941 until 1945, and another 60 000 tons of ore were extracted. (Detailed mine plans and sections were produced by Groves, 1952). Apart from these two major excavations, minor trials for manganese were made along the southern tip of Trwyn y Penrhyn [1887 2525].

The manganese ore in this district is generally a replacement of ferruginous debris flows bearing a mixture of basement-derived clastic materials, as well as fresh volcanoclastic debris, intraclasts, pisoids, ooids and oncoids.

The ore is very strongly magnetic, due to the presence of the manganese-iron oxide jacobsite; the intensity of magnetisation has been shown to be approximately proportional to grade (Brown and Evans, 1989). On this basis, the drift-covered ground between the Nant and Benallt mines, and the Mynydd Rhiw area north of Benallt, were investigated by a detailed magnetometer survey (Brown and Evans, 1989). The results were largely negative. The subeconomic mineralisation proved between Nant and Benallt by drilling in 1942 clearly does not develop laterally into ore grade, and there is no evidence of any other substantial manganese ore body close to surface in the area. Drilling to test two localised magnetic anomalies, near Tyddyn Meirion and on Mynydd Rhiw, proved strongly magnetic stratabound ironstones with a maximum dimension of only a few tens of metres.

Benallt Mine

The manganiferous stata at Benallt Mine lie beneath the 'Clip Lava' and above the 'Footwall Sill', a dolerite correlated with the thick Gallt y Mor intrusion at the coast. In detail, however, the geology of this mining belt is complicated by intense deformation : thus 'the mudstones of the mining belt were intensely folded, overfolded, faulted and thrusted to form what is known as imbricate structure' (Groves, 1952, p.303). Steeply south-east-dipping (c.60°), lenticular, fault-bounded ore bodies occur *en échelon* or in closely spaced groups within the mudstone host rock. The ore is a hard, compact rock with a 25 to 35 per cent manganese content, and is composed mostly of a complex mixture of manganese silicates (mostly pennantite), with subsidiary manganese and iron oxides. Other manganese minerals occurring with the pennantite include tephroite, granophyllite (Campbell-Smith and Bannister, 1948), spessartine, alleghenyite, pyrochlorite, jacobsite and rhodonite. Benallt Mine has provided the first British examples of pennantite, banalsite and cymrite (Campbell-Smith et al., 1944).

The sediments associated with the ore display stromatolitic, pisolitic and oolitic textures, and are commonly phosphatic and 'chamositic'. In particular, a gradation can be traced from 'chamositic' mudstone (and in places sandstone) into manganese ore, and it has therefore been concluded that the ore was produced by the metasomatisation of the 'chamositic' sediments (e.g. Groves, 1952). Ferruginous siltstones and sandstones lying beneath the dolerite sill exposed on the south-west side of Nant y Gadwen may represent unmetasomatised equivalents of the host rock.

Nant Mine

As with the Benallt Mine, the ore body in Nant is entirely bounded by faults, described as 'thrusts' by Groves (1952). Another similarity between the two mines is the presence of basic igneous bodies both above and below the ore. However, the upper igneous body, exposed in south-west Nant y Gadwen, is an intrusive dolerite sill rather than a pillow lava, as at Benallt. Assuming that this sill is indeed the same body as the 'Clip Lava' (see p.43), then it must be highly variable in form, possessing a pillowed top and doleritic base.

Although the geological setting of both mines is similar, the ore won from each was rather different. The main difference was that the Nant ore was much richer in manganese carbonate. The ore protolith was considered by Woodland (1956) to be an albitic tuffaceous rock in which the albites have been metasomatically replaced by rhodochrosite, sericitic mica, epidote and chlorite in a cryptocrystalline, hematitised matrix displaying stromatolitic and pisolitic textures. Thus, whereas the dominant protolith for the ore in Nant Mine was tuffaceous (presumably similar to the unmetasomatised tuffs exposed in the valley), that at Benallt was a 'chamositic' mudstone.

REFERENCES

Most of the references listed below are held in the Library of the British Geological Survey at Keyworth, Nottingham. Copies of the references can be purchased from the Library subject to the current copyright legislation.

ALLEN J R L. 1982. Late Pleistocene (Devensian) glaciofluvial outwash at Banc-y-Warren near Cardigan (West Wales). *Geological Journal*, Vol. 17, 31–47.

BAKER, J W. 1969. Correlation problems of unmetamorphosed Pre-Cambrian rocks in Wales and southeast Ireland. *Geological Magazine*, Vol. 106, 246–259.

BATTIAU-QUENNEY, Y. 1980. Contribution a l'étude géomorphologique du Massif Gallois. These, l'Universite de Bretagne Occidentale.

— 1984. The pre-glacial evolution of Wales. *Earth Surface Processes and Landforms*, Vol. 9, 229–252.

— 1987. Tertiary inheritance in the present landscape of the British Isles. 979–989 in *International geomorphology 1986, Part II*. GARDNER, V (editor). (Chichester: John Wiley & Sons.)

— and SAUCEROTT, M. 1985. Paleosol preglaciaire de la carriere de Ballyegan (Kerry, Irlande). *Hommes et Terres de Nord*, Vol. 3, 234–237.

BEAUJEAU-GARNIER, J. 1975. Variety of the natural elements. HOUSTON, J M (editor). 4–19 (London: Longman.)

BECKET, S C. 1977. The Bog Roos. 42–45 in *X INQUA Congress Excursion Guide; Yorkshire and Lincolnshire*. CATT, J A (editor). (Norwich: Geo Abstracts.)

BECKINSALE, R D, EVANS, J A, THORPE, R S, GIBBONS, W, and HARMON, R S. 1984. Rb-Sr whole-rock isochron ages, $\delta^{18}O$ values and geochemical data for the Sarn Igneous Complex and the Parwyd gneisses of the Mona Complex of Llŷn, N Wales. *Journal of the Geological Society of London*, Vol. 141, 701–709.

BECKLY, A J. 1985. The Arenig Series in North Wales. Unpublished PhD thesis, University of London.

— 1987. Basin development in North Wales during the Arenig. *Geological Journal*, Vol. 22, 19–30.

— 1988. The stratigraphy of the Arenig series in the Aberdaron to Sarn area, western Llŷn, North Wales. *Geological Journal*, Vol. 23, 321–337.

BELBIN, S. 1985. Long-term landform development in northwest England: the application of the planation concept. 37–58 in *The geomorphology of north-west England*. JOHNSON, R H (editor). (Manchester: Manchester University Press.)

BLAKE, J F. 1888. On the Monian system of rocks. *Quarterly Journal of the Geological Society of London*, Vol. 44, 463–546.

BLOXHAM, T N, AND DIRK, M H J. 1988. The petrology and geochemistry of the St David's granophyre and the Cwm Bach rhyolite, Pembrokeshire, Dyfed. *Mineralogical Magazine*, Vol. 52, 563–575.

BONNEY, T G, 1881. On the serpentine and associated rocks of Anglesey; with a note on the so-called serpentine of Porth Dinlleyn (Caernarvonshire). *Quarterly Journal of the Geological Society of London*, Vol. 37, 48–50.

— 1885. On the so-called diorite of Little Knott (Cumberland), with further remarks on the occurrence of picrites in Wales. *Quarterly Journal of the Geological Society of London*, Vol. 41, 511–521.

BOULTON, G S. 1972. The role of thermal regime in glacial sedimentation. 1–19 in Polar geomorphology. PRICE R J and SUGDEN, D E (editors). *Institute of British Geographers Special Publication*, No. 4.

— 1977. A multiple till sequence formed by a Late Devensian Welsh ice-cap; Glanllynau, Gwynedd. *Cambria*, Vol. 1, 10–31.

— and JONES, A S. 1979. The stability of temperate ice caps and ice sheets resting on beds of deformable sediment. *Journal of Glaciology*, Vol. 24, 29–43.

BOUMA, A H. 1962. *Sedimentology of some flysch deposits: a graphic approach to facies interpretation*. (Amsterdam: Elsevier.)

BOWEN, D G. 1969. The Pleistocene history of the Bristol Channel. *Proceedings of the Ussher Society*, Vol. 2, 86.

— 1973. The Pleistocene succession in the Irish Sea. *Proceedings of the Geologists' Association*, Vol. 84, 249–272.

— 1977. The coast of Wales. *In* The Quaternary history of the Irish Sea. KIDSON, C, and TOOLEY, M J (editors). *Geological Journal Special Issue*, No. 7.

— ROSE, J, MCCABE, A M, and SUTHERLAND, D G. 1986. Correlation of Quaternary glaciations in England, Ireland, Scotland and Wales. *Quaternary Science Reviews*, Vol. 5, 299–340.

— and SYKES, G A. 1988. Correlation of marine events and glaciations on the northeast Atlantic margin. *Philosophical Transactions of the Royal Society*, B, Vol. 318, 619–635.

— SYKES, G A, REEVES, A, MILLER, G H, ANDREWS, G T, BREW, J S and HARE, P E. 1985. Amino acid geochronology of raised beaches in southwest Britain. *Quaternary Science Reviews*, Vol. 4, 279–318.

BRADSHAW, M. 1982. Process, time and the physical landscape; geomorphology today. *Geography*, Vol. 67, 15–28.

BREMER, H. 1980. Landform development in the humid tropics; German geomorphological research. *Zeitschrift für Geomorphologie*, Vol. 36, 162–175.

BROWN, E H. 1952. The 600 foot platform in Wales. 304 in *Proceedings of the 8th General Assembly and 17th International Congress of the International Geographical Union*.

— 1960. *The relief and drainage of Wales*. (Cardiff: University of Wales Press.)

BROWN, M, POWER, G M, TOPLEY, C G and D'LEMOS, R S. 1990. Cadomian magmatism in the North Armorican massif. 245–259 *in* The Cadomian Orogeny. D'LEMOS, R S, STRACHAN, R A, and TOPLEY, C G (editors). *Special Publication of the Geological Society of London*, No. 51.

BROWN, M J, and EVANS, A D. 1989. Geophysical and geochemical investigations of the manganese deposits of Rhiw, western Llŷn, North Wales. *British Geological Survey Technical Report*, WF/89/14.

BRUNSDEN, D. 1980. Applicable models of longterm landform evolution. *Zeitschrift für Geomorphologie, supplementband*, Vol. 36, 16–26.

BÜDEL, J. 1957. Die "Doppelten Einebnungsflachen" in den feuchten Tropen. *Zeitschrift für Geomorphologie*, Vol. 1, 201–288.

— 1979. Reliefgenerationem und klimageschichte in Mitteleuropa. *Zeitschrift für Geomorphologie, Supplementband*, Vol. 33, 1–15.

CAMPBELL-SMITH, W, and BANNISTER, F A. 1948. Granophyllite from the Benallt mine, Rhiw, Caernarvonshire. *Mineralogical Magazine*, Vol. 28, 343.

— and CLARINGBULL, G F. 1947. Pyrophanite from the Bennallt Mine, Rhiw, Caernarvonshire. *Mineralogical Magazine*, Vol. 28, 108.

— BANNISTER, F A, and HEY, M H. 1944. Banalsite, a new barium feldspar from Wales. With an appendix by A W Groves. *Mineralogical Magazine*, Vol. 27, 33.

— — — 1946. Pennantite, a new manganese-rich chlorite from Benallt Mine, Rhiw, Caernarvonshire. *Mineralogical Magazine*, Vol. 27, 217.

— — — 1949. Cymrite, a new barium mineral from the Benallt mine, Rhiw, Caernarvonshire. *Mineralogical Magazine*, Vol. 28, 676.

CAMPBELL, S, and BOWEN, D Q. 1990. *Geological conservation review: Quaternary of Wales.* (Nature Conservancy Council.)

CATTERMOLE, P J. 1969. A preliminary geochemical study of the Mynydd Penarfynydd Intrusion, Rhiw Igneous Complex, south-west Lleyn. 435–446 in *Pre-Cambrian and Lower Palaeozoic rocks of Wales*. WOOD, A (editor). (Cardiff: University of Wales Press.)

— 1976. The crystallisation and differentiation of a layered intrusion of hydrated alkali olivine-basalt parentage at Rhiw, North Wales. *Geological Journal*, Vol. 11, 45–70.

— and FUGE, R. 1969. The abundance and distribution of fluorine and chlorine in a layered intrusion at Rhiw, North Wales. *Geochimica and Cosmochimica Acta*, Vol. 33, 1295.

COLHOUN, E A, and MCCABE, A M. 1973. Pleistocene glacial, glaciomarine and associated deposits of Mell and Tullyallen townlands, near Drogheda, eastern Ireland. *Proceedings of the Royal Irish Academy*, Vol. 73B, 165–206.

CONEY, P J, JONES, D L, and MONGER, J W H. 1980. Cordilleran suspect terranes. *Nature, London*, Vol. 188, 329–333.

CONNELL, E R, and HALL, A M. 1984. Kirkhill Quarry. 80–81 in *Buchan Field Guide*. HALL, A M (editor). (Cambridge: Quaternary Research Association.)

COOPE, G R, and BROPHY, J A. 1972. Late-glacial environmental changes indicated by a coleopteran succession from North Wales. *Boreas*, Vol. 1, 97–142.

— and JOACHIM, M J. 1980. Lateglacial environment changes interpreted from fossil coleoptera from St Bees, Cumbria, northwest England. In *Studies in the Lateglacial of northwest Europe*. LOW, J J, GRAY, J M, and ROBINSON, J E (editors). (Oxford: Pergamon Press.)

CRIMES, T P. 1969. Trace fossils from the Cambro-Ordovician rocks of North Wales and their stratigraphic significance. *Geological Journal*, Vol. 6, 333–338.

— 1970. A facies analysis of the Arenig of western Lleyn, North Wales. *Proceedings of the Geologists' Association*, Vol. 81, 221–239.

DACKOMBE, R V, and THOMAS, G S P. 1985. *Field guide to the Quaternary of the Isle of Man.* (Cambridge: Quaternary Research Association.)

DAWSON, A G, MATTHEWS, J A, and SHAKESBY, R A. 1987. Rock platform erosion on periglacial shores: a modern analogue for Pleistocene rock platforms in Britain. In *Periglacial processes and landforms in Britain and Ireland*. BOARDMAN, J (editor). (Cambridge University Press.)

DEWEY, H, and DINES, H G. 1923. Tungsten and manganese ores. *Memoir of the Geological Survey, Mineral Resources*, Vol. 1.

ELLES, G L. 1904. Graptolite zones in the Arenig rocks of Wales. *Geological Magazine*, Vol. 1, 199–211.

ELSDEN, J V. 1888. Notes on the igneous rocks of the Lleyn promontory. *Geological Magazine*, Vol. 5, 303–308.

EMBLETON, C. 1964. The planation surfaces of Arfon and adjacent parts of Anglesey: a re-examination of their age and origin. *Transactions of the Institute of British Geographers*, Vol. 35, 17–26.

EYLES, C H, and EYLES, N. 1984. Glaciomarine sediments of the Isle of Man as a key to Late Pleistocene stratigraphic investigations in the Irish Sea Basin. *Geology*, Vol. 12, 359–364.

— — and MCCABE, A N. 1985. Glaciomarine sediments of the Isle of Man as a key to Late Pleistocene stratigraphic investigations in the Irish Sea Basin: a reply. *Geology*, Vol. 13, 446–447.

— and MCCABE, A M. 1989a. Glaciomarine deposits of the Irish Sea Basin: the role of glaciosostatic disequilibrium. In *Glacial deposits of the British Isles*. EHLERS, J, GIBBARD, P, and ROSE, J (editors). (Rotterdam: Balkema.)

— — 1989b. Glaciomarine facies within subglacial tunnel valleys: the sedimentary record of glacio-isostatic downwarping in the Irish Sea Basin. *Sedimentology*, Vol. 36, 431–448.

— — 1989c. The Late Devensian (<22 000 YBP) Irish Sea Basin: the sedimentary record of a collapsed ice sheet margin. *Quaternary Science Reviews*, Vol. 8, 307–351.

— — and MIALL, A D. 1983. Lithofacies types and vertical profile models; an alternative approach to the description and environmental interpretation of glacial diamict and diamictite sequences. *Sedimentology*, Vol. 30, 393–410.

FAIRBRIDGE, R. 1977. Rates of sea-ice erosion of Quaternary littoral platforms. *Studia Geological Polonica*, Vol. 52, 135–142.

FITCH, F J, MILLER, J A, and MENEISY, M J. 1963. Geochronological investigations on rocks from North Wales. *Nature, London*, Vol. 199, 449–451.

FORTEY, R A, BECKLY, A J, and RUSHTON, W A. 1990. International correlation of the base of the Llanvirn Series Ordovician System. *Newsletters on Stratigraphy*, Vol. 22, 119–142

— and OWENS, R M. 1987. The Arenig Series in South Wales. *Bulletin of the British Museum (Natural History)*, Vol. 41, 69–307.

GIBBONS, W. 1980. The geology of the Mona Complex of the Lleyn Peninsula and Bardsey Island, North Wales. Unpublished PhD thesis, Portsmouth Polytechnic.

— 1981. Glaucophanic amphibole in the Monian shear zone on the mainland of N Wales. *Journal of the Geological Society of London*, Vol. 138, 139–143.

— 1983a. The Monian 'Penmynydd Zone of Metamorphism' in Llyn, North Wales. *Geological Journal*, Vol. 18, 21–41.

— 1983b. Stratigraphy, subduction and strike-slip faulting in the Mona Complex of North Wales—a review. *Proceedings of the Geologists' Association*, Vol. 94, 147–163.

— 1989a. Basement-cover relationships around Aberdaron, Wales, UK: the fault reactivated northwestern margin of the Welsh Basin. *Geological Magazine*, Vol. 126, 363–372.

— 1989b. Suspect terrane definition in Anglesey, North Wales. 59–66 in Terranes in the Circum-Atlantic Palaeozoic Orogens. DALLMEYER, R D (editor). *Geological Society of America, Special Paper*, No 230.

— 1990a. Transcurrent ductile shear zones and the dispersal of the Avalon superterrane. 407–423 in The Cadomian Orogeny. D'LEMOS, R S, STRACHAN, R A, and TOPLEY, C G (editors). *Special Publication of the Geological Society of London*, No. 51, 407–423.

— 1990b. Pre-Arenig terranes of northwest Wales. 28–43 in *Avalonian and Cadomian geology of the North Atlantic*. STRACHAN, R A, and TAYLOR, G K (editors). (Glasgow & London: Blackie.)

— 1987. Menai Strait fault system: an early Caledonian terrane boundary in north Wales. *Geology*, Vol. 15, 744–747.

— and BALL, M J. 1991. Discussion on Monian Supergroup stratigraphy in northwest Wales. *Journal of the Geological Society of London*, Vol. 148, 5–8.

GOODCHILD, J G. 1875. The glacial geology of the Eden valley and the western part of the Yorkshire-Dale district. *Quarterley Journal of the Geological Society of London*, Vol. 31, 55–99.

GREENLY, E. 1919. The geology of Anglesey. *Memoir of the Geological Survey of Great Britain*, 2 Vols. 980 pp.

— 1938. The age of the mountains of Snowdonia. *Quarterly Journal of the Geological Society of London*, Vol. 94, 117–122.

GROVES, A W. 1947. Results of magnetometric survey at Benallt Manganese Mine, Rhiw, Caernarvonshire. *Bulletin of the Institute of Mining and Metallurgy*, No. 484, 1.

— 1952. Wartime investigations into the hematite and manganese resources of Great Britain and Northern Ireland. *Permanent Records of Research and Development, Ministry of Supply, London*, No. 20, 703.

GUILCHER, A. 1969. Le Quaternaire littoral et sous-marin dans l'Atlantique (Côtes Françaises). 33–41 in Françaises sur le Quaternaire. VIII Congress INQUA.

HALL, A M. 1985. Cenozoic weathering covers in Buchan, Scotland, and their significance. *Nature, London*, Vol. 315, 392–395.

— 1986. Deep weathering patterns in northeast Scotland and their geomorphological significance. *Zeitschrift für Geomorphologie*, Vol. 30, 407–422.

— 1987. Weathering and relief development in Buchan, Scotland. 991–1005 in International Geomorphology 1986, Part II. GARDNER, V (editor). (Chichester: Wiley.)

HARKER, A. 1888. The eruptive rocks of Sarn. *Quarterly Journal of the Geological Society of London*, Vol. 44, 442–461.

— 1889. *The Bala volcanic series of Caernarvonshire and associated rocks (Sedgwick Prize Essay)*. (Cambridge.)

HARRIS, C. 1989. Glacially deformed bedrock at Wylfa Head, Anglesey North Wales. 31–42 in *Quaternary Engineering Geology*: preprints of papers for the 25th annual conference of the engineering group of the Geological Society.

HARRISON, R K. 1971. The petrology of the Upper Triassic rocks in the Llanbedr (Mocras Form) borehole. *Report of the Institute of Geological Sciences*, No. 71/18, 37–72.

HAWKINS, T R W. 1965. A note on rhythmic layering in hornblende-bearing basic and ultrabasic intrusions near Rhiw, Caernarvonshire. *Geological Magazine*, Vol. 100, 222.

— 1967. The geology of the Lower Palaeozoic rocks between Aberdaron and Rhiw, Caernarvonshire, North Wales. Unpublished PhD thesis, University of London.

— 1970. Hornblende gabbros and picrites at Rhiw, Caernarvonshire. *Geological Journal*, Vol. 7, 1–24.

— 1983. Structure of the Ordovician succession around Aberdaron, southwest Llyn, North Wales. *Geological Journal*, Vol. 18, 169–181.

HICKS, H. 1878. On some new Pre-Cambrian areas in Wales. *Geological Magazine*, Vol. 5, 460–461.

— 1879. On the Pre-Cambrian (Dimetian, Arvonian and Pebidian) rocks in Caernarvonshire and Anglesey (with appendix by T G Bonney). *Quarterly Journal of the Geological Society of London*, Vol. 35, 295–308.

HOPLEY, D. 1963. The coastal geomorphology of Anglesey. Unpublished MA thesis, University of Manchester.

JACKSON, D E. 1962. Graptolite zones in the Skiddaw Group in Cumberland, England. *Journal of Paleontology*, Vol. 36, 300–313.

JEHU, T J. 1909. The glacial deposits of Western Caernarvonshire. *Transactions of the Royal Society of Edinburgh*, Vol. 47, 17–56.

JOHN, B A. 1970. 229–265 in *The glaciations of Wales and adjoining regions*. LEWIS, C A (editor). (London: Longman.)

JONES, R L. 1977. Late Devensian deposits from Kildale, northeast Yorkshire. *Proceedings of the Yorkshire Geological Society*, Vol. 41, 185–188.

KEEN, D H. 1978. The Pleistocene deposits of the Channel Islands. *Report of the Institute of Geological Sciences*, No. 78/28, 1–14.

— JONES, R L, and ROBINSON, J E. 1984. A Late Devensian fauna and flora from Kildale, northeast Yorkshire. *Proceedings of the Yorkshire Geological Society*, Vol. 441, 385–397.

KENNEDY, R J. 1989. Ordovician (Lanvirn) trilobites from SW Wales. *Monograph of the Palaeontographical Society*, No. 576.

KIDSON, C. 1977. Some problems of the Quaternary of the Irish Sea. *In* The Quaternary history of the Irish Sea. KIDSON, C, and TOOLEY, M J (editors). *Geological Journal Special Issue*, No. 7.

KING, C A M. 1963. Some problems concerning marine planation and the formation of erosion surfaces. *Transactions of the Institute of British Geographers*, Vol. 33, 29–43.

KING, L C. 1953. Canons of landscape evolution. *Bulletin of the Geological Society of America*, Vol. 64, 721–752.

— 1957. The uniformitarian nature of hillslopes. *Transactions of the Edinburgh Geological Society*, Vol. 17, 81–102.

KROONENBERG, S B, and MELITZ, P J. 1983. Summit levels, bedrock control and the etchplain concept in the basement of Surinam. *Geologie en Mijnbouw*, Vol. 62, 389–400.

LAMPLUGH, G W. 1903. The geology of the Isle of Man. *Memoir of the Geological Survey of Great Britain*.

LEAT, P T, and THORPE, R S. 1989. Snowdon basalts and the cessation of Caledonian subduction by the Longvillian. *Journal of the Geological Society of London*, Vol. 146, 965–970.

LEWIS, C. 1894. *The glacial geology of Britain and Ireland*.

LINDMAR-BERGSTRÖM, K. 1982. Pre-Quaternary geomorphological evolution in southern Fennoscandia. *Sveriges Geologiska Unsersokning*, Sevie C, No. 785, 1-202.

LOWE, J J, and WALKER, M J C. 1984. *Reconstructing Quaternary environments*. (London: Longman.)

MATLEY, C A. 1902. The Arenig rocks near Aberdaron. *Geological Magazine*, Vol. 9, 118–122.

— 1913. The geology of Bardsey Island. *Quarterly Journal of the Geological Society of London*, Vol. 69, 514–533.

— 1925. An outline of the geology of south-western Lleyn. *Geological Magazine*, Vol. 62, 129–132.

— 1928. The Pre-Cambrian complex and associated rocks of south-western Lleyn (Carnarvonshire). *Quarterly Journal of the Geological Society of London*, Vol. 84, 440–501.

— 1932. The geology of the country around Mynydd Rhiw and Sarn, SW Lleyn, Caernarvonshire. *Quarterly Journal of the Geological Society of London*, Vol. 88, 238–273.

— 1939. Summer field meeting in western Lleyn. *Proceedings of the Geologists' Association*, Vol. 50, 83–100.

— and SMITH, B. 1936. The age of the Sarn Granite. *Quarterly Journal of the Geological Society of London*, Vol. 92, 188–200.

MATTHEWS, J A, DAWSON, A G, and SHAKESBY, R A. 1986. Lake shoreline development, frost weathering and rock platform erosion in an alpine periglacial environment, Jotunheimen, southern Norway. *Boreas*, Vol. 15, 33–50.

MCCABE, A M. 1986a. Late Pleistocene tidewater glaciers and glacial marine sequences from north County Mayo, Republic of Ireland. *Journal of Quaternary Science*, Vol. 2, 73–84.

— 1986b. Glaciomarine facies deposited by retreating tidewater glaciers. An example from the Late Pleistocene of Northern Ireland. *Journal of Sedimentary Petrology*, Vol. 56, 886–894.

— 1987a. Quaternary deposits and glacial straigraphy in Ireland: a review. *Quaternary Science Reviews*, Vol. 6, 259–300.

— 1987b. Glacial facies deposited by retreating tidewater glaciers: an example from the Late Pleistocene of Northern Ireland. *Journal of Sedimentary Petrology*, Vol. 56, 880–894.

— DARDIS, G F, and HANVEY, P M. 1984. Sedimentology of a Late Pleistocene submarine-moraine complex, County Down, Northern Ireland. *Journal of Sedimentology*, Vol. 54, 716–730.

— — — 1987. Sedimentation at the margins of a Late Pleistocene ice-lobe terminating in shallow marine environments, Dundalk Bay, eastern Ireland. *Sedimentology*, Vol. 34, 473–493.

— and EYLES, N. 1988. Sedimentology of an ice-contact delta, Carey Valley, Northern Ireland. *Sedimentary Geology*, Vol. 39, 1-14.

— HAYNES, J R, and MACMILLAN, N. 1986. Late Pleistocene tidewater glaciers and glaciomarine sequences from north County Mayo, Republic of Ireland. *Journal of Quaternary Science*, Vol. 1, 73–84.

— and HIRONS, K R. 1986. Field guide to the Quaternary deposits in southeastern Ulster. *Quaternary Research Association*. (Cambridge.)

MCCARROLL, D. 1991. Ice directions in western Lleyn and the status of the Gwynedd readvance of the last Irish Sea glacier. *Geological Journal*, Vol. 26, 137–143.

MCWILLIAMS, M D, and HOWELL, D G. 1982. Exotic terranes of western California. *Nature, London*, Vol. 297, 215–217.

MITCHELL, G F. 1960. The Pleistocene history of the Irish Sea. *Advancement of Science*, Vol. 17, 313–325.

— 1972. The Pleistocene history of the Irish Sea: second approximation. *Scientific Proceedings of the Royal Dublin Society*, Vol. 4A, 181–199.

— 1972. The Pleistocene history of the Irish Sea: second approximation. *Scientific Proceedings of the Royal Dublin Society*, Vol. 4A, 181–199.

MOLYNEUX, S G. 1987. Acritarchs and Chitinozoa from the Arenig Series of south-west Wales. *Bulletin of the British Museum (Natural History)*, Vol. 41, 309–364.

MOTTERSHEAD, D N. 1977. The Quaternary evolution of the south coast of England. *In* The Quaternary history of the Irish Sea. KIDSON, C, and TOOLEY, M J (editors). *Geological Journal Special Issue*, No. 7.

OLLIER, C. 1981. *Tectonics and landforms*. (Longman: London.)

PAGE, B M, and SUPPE, J. 1981. The Pliocene Lichi Mélange of Taiwan: its plate tectonic and olistostromal origin. *American Journal of Science*, Vol. 281, 193–227.

PEARCE, J A, HARRIS, B W, and TINDLE, A G. 1984. Trace element discrimination diagrams for the tectonic interpretation of granitic rocks. *Journal of Petrology*, Vol. 25, 956–983.

PECCERILLO, A, and TAYLOR, S R. 1976. Geochemistry of Eocene calc-alkaline volcanic rocks from the Kastamonou area, northern Turkey. *Contributions to Mineralogy and Petrology*, Vol. 58, 63–81.

PENNY, L F, COOPE, G R, and CATT, J A. 1969. Age and insect fauna of the Dimlington silts, east Yorkshire. *Nature, London*, Vol. 224, 65–67.

PINCHEMEL, P. 1969. *France: a geographical survey*. (London: Bell.)

RAISIN, C A. 1893. Variolite of the Lleyn and associated volcanic rocks. *Quarterly Journal of the Geological Society of London*, Vol. 49, 145–165.

RAMSAY, A C. 1866. The geology of North Wales. *Memoir of the Geological Survey of Great Britain*.

— 1881. The geology of North Wales. *Memoir of the Geological Survey of Great Britain* (2nd edition).

ROSE, J. 1985. The Dimlington stadial/Dimlington chronozone: a proposal for naming the main glacial episode of the Late Devensian in Britain. *Boreas*, Vol. 14, 225–230.

— 1987. Status of the Wolstonian glaciation in the British Quaternary. *Quaternary Newsletter*, No. 53, 1–9.

ROWLANDS, P H. 1971. Radiocarbon evidence of the age of an Irish Sea glaciation in Vale of Clwyd. *Nature, London*, Vol. 230, 9–11.

SAUNDERS, G E. 1968a. Fabric analysis of the ground moraine deposits of the Lleyn Peninsula of southwest Caernarvonshire. *Geological Journal*, Vol. 6, 105–188.

— 1968b. A reappraisal of glacial drainage phenomena in the Lleyn Peninsula. *Proceedings of the Geologists' Association*, Vol. 79, 305–324.

SCHUSTER, D C. 1979. The Gwna Mélange, a late Precambrian olistostromal sequence, North Wales, United Kingdom (abstract). *Bulletin of the Association of Petroleum Geologists*, No. 63, 523.

— 1980. The nature and origin of the late Precambrian Gwna Mélange, North Wales, United Kingdom. Unpublished PhD thesis, University of Illinois at Urbana-Champaign, USA.

SHACKLETON, N J. 1987. Oxygen isotopes, ice volume and sea level. *Quaternary Science Reviews*, Vol. 6, 183–190.

— and 16 others. 1984. Oxygen isotope calibration of the onset of ice-rafting and history of glaciation in the North Atlantic region. *Nature, London*, No. 307, 620–623.

— and OPDYKE, N D. 1978. Oxygen isotope and palaeomagnetic stratigraphy of Pacific core V28–239, Late Pliocene to Late Holocene. *Geological Society of America Memoirs*, No. 145, 449–464.

SHACKLETON, R M. 1956. Notes on the structure and relations of the Pre-Cambrian and Ordovician rocks of south-western Lleyn (Carnarvonshire). *Geological Journal*, Vol. 1, 400–409.

SHARPE, D. 1846. Contributions to the geology of North Wales. *Quarterly Journal of the Geological Society of London*, Vol. 2, 283–314.

SISSONS, J B. 1981. British shore platforms and ice-sheets. *Nature, London*, Vol. 291, 473–475.

— and DAWSON, A G. 1981. Former sea levels and ice limits in part of Wester Ross, northwest Scotland. *Proceedings of the Geologists' Association*, Vol. 92, 115–124.

STEPHENS, N. 1970. The West Country and Southern Ireland. 267–314 *in* The glaciations of Wales and adjoining regions. LEWIS, C A (editor). (London: Longman.)

— 1980. 'Introduction' In *Field handbook: west Cornwall meeting.* SIMMS, P C, and STEPHENS, N (editors). Quaternary Research Association (Plymouth: Plymouth Polytechnic.)

STRAHAN, A, and CANTRILL, T C. 1904. The geology of the country around Bridgend. *Memoir of the Geological Survey of Great Britain.*

STREIKEISEN, A L. 1976. To each plutonic rock its proper name. *Earth Science Reviews*, Vol. 12, 1–33.

SYNGE, F M. 1964. The glacial succession in west Caernarvonshire. *Proceedings of the Geologists' Association*, Vol. 75, 431–444.

TAWNEY, E B. 1880. Woodwardian laboratory notes. North Wales rocks. *Geological Magazine*, Vol. 7, 207–215, 452–458.

— 1883. Woodwardian laboratory notes: North Wales rocks. *Geological Magazine*, Vol. 10, 65–71.

TEALL, J J H. 1888. *British petrography.* (London.)

THOMAS, G S P. 1976. The Quaternary stratigraphy of the Isle of Man. *Proceedings of the Geologists' Association*, Vol. 87, 307–323.

— 1977. The Quaternary of the Isle of Man. 155–178 *in* Quaternary history of the Irish Sea. KIDSON, C, and TOOLEY, M J (editors). *Geological Journal Special Issue*, No. 7.

— 1984. A late Devensian glaciolacustrine fan-delta at Rhosesmor, Clwyd, North Wales. *Geological Journal*, Vol. 19, 125–141.

— 1985. The Quaternary of the northern Irish Sea Basin. 143–158 in *The geomorphology of northwest England.* JOHNSON, R H (editor). (Manchester University Press.)

— 1983. The Quaternary stratigraphy between Blackwater Harbour and Tinnaberna, County Wexford. *Journal of Earth Sciences, Royal Dublin Society*, Vol. 5, 121–134.

— CONNAUGHTON, M, and DACKOMBE, R V. 1985. Facies variations in a Late Pleistocene supraglacial outwash sandur from the Isle of Man. *Geological Journal*, Vol. 22, 193–213.

— and KERR, P. 1987. The stratigraphy, sedimentology and palaeontology of the Pleistocene knocknasillogge member, County Wexford, Ireland. *Geological Journal*, Vol. 22, 67–82.

— and SUMMERS, A J. 1982. Drop-stone and allied structures from Pleistocene waterlain till at Ely House, County Wexford. *Journal of Earth Sciences, Royal Dublin Society*, Vol. 4, 109–122.

— 1984. Glaciodynamic structures from the Blackwater formation, County Wexford, Ireland. *Boreas*, Vol. 13, 5–12.

THOMAS M F. 1974. T*ropical geomorphology: a study of weathering and landform development in warm climates.* (London: Macmillan.)

— and SUMMERFIELD, M A. 1987. Long-term landform development: key themes and research problems. 935–956 in *International Geomorphology* 1986, Part II. GARDNER, U (editor). (Chichester: Wiley)

THORPE, M B. 1967. Joint patterns and landforms in the Jarawa granite massif, northern Nigeria. 65–84 in *Essays in Geography.* STEEL, R W, and LAWTON, R (editors). (London: Longman.)

TOPLEY, C G, BROWN, M, D'LEMOS, R S, POWER, G M, and ROACH, R A. 1990. The Northern Igneous Complex of Guernsey, Channel Islands. 245–249 in The Cadomian Orogeny. D'LEMOS, R S, STRACHAN, R A, and TOPLEY, C G (editors). *Special Publication of the Geological Society of London*, No. 51.

TRICART, J. 1972. Landforms of the humid tropics: forest and savanna. (London: Longman.)

TRIMMER, J. 1831. On the diluvial deposits of Caernarvonshire between the Snowdon chain of hills and the Menai Straits. *Proceedings of the Geological Society of London*, Vol. 1, 331–332.

TWIDALE, C R. 1976. On the survival of palaeoforms. *American Journal of Science*, Vol. 276, 77–95.

— 1981. Pediments, peneplains and ultiplains. *Revue de Geomorphologie Dynamique*, Vol. 32, 1–35.

WALSH, P T, ATKINSON, K, BOULTER, M C, and SHAKESBY, R A. 1987. The Oligocene and Miocene outliers of west Cornwall and their bearing on the geomorphological evolution of Oldland Britain. *Philosophical Transactions of the Royal Society of London*, A, Vol. 323, 211–245.

WATERS, R S. 1957. Differential weathering on oldlands. *Geographical Journal*, Vol. 123, 503–509.

WAYLAND, E J. 1934. Peneplains and some other erosional platforms. *Bulletin of the Geological Survey, Uganda, Annual Report*, No. 74, 366.

WHITTOW, J B. 1957. The Lleyn Peninsula, North Wales. A geomorphological study. Unpublished PhD thesis, University of Reading.

— 1960. Some comments on the raised beach platform of southwest Caernarvonshire and on an unrecorded raised beach at Porth Neigwl, North Wales. *Proceedings of the Geologists' Association*, Vol. 71, 31–39.

— 1965. The interglacial and postglacial strandlines of North Wales. 94–117 in *Essays in geography for Austin Miller.* WHITTOW, J B, and WOOD, P D (editors). (Reading.)

— and BALL, D F. 1970. North-west Wales. 21–58 in *The glaciations of Wales and adjoining areas.* LEWIS, C A (editor). (London: Longman.)

WINTLE, A G, and CATT, J A. 1985. Thermoluminescence dating of the Dimlington Stadial deposits in eastern England. *Boreas*, Vol. 14, 231–234.

WOOD, D S. 1974. Ophiolites, mélanges, blueschists and ignimbrites: early Caledonian subduction in Wales?. 334–344 in *Modern and ancient geosynclinal sedimentation.* DOTT, R H, Jr., and SHAVER, R H (editors). *Special Publication of the Society of Economic Palaeontologists and Mineralogists*, No. 19.

— and SCHUSTER, D C. 1978. The nature of mélanges; criteria for recognition of their origin with reference to the "type" mélange in Wales. *Geological Society of America Abstracts with Programs*, Vol. 10: 519.

WOODLAND, A W. 1938. Some petrological studies in the Harlech Grit Series of Merionethshire. *Geological Magazine,* Vol. 75, 366, 441 and 529.

— 1939a. The petrography and petrology of the Lower Cambrian manganese ore of west Merionethshire. *Quarterly Journal of the Geological Society of London,* Vol. 95, 1.

— 1939b. The petrography and petrology of the manganese ore of the Rhiw district (Carnarvonshire). *Proceedings of the Geologists' Association,* Vol. 50, 205–222.

— 1956. The manganese deposits of Great Britain. 197–218 in *Symposium sobre yacimientos demanganeso,* Volume 5, Europe. REYNA, J G (editor).

YOUNG, T P. 1991. A revision of the age of the Hen-dy-capel Ooidal Ironstone (Ordovician), Llanengan, N. Wales. *Geological Journal,* Vol. 26, 317–327.

— GIBBONS, W, and MCCARROLL, D. In press. Geology of the country around Pwllheli. *Memoir of the British Geological Survey* (Sheet 134, England and Wales).

APPENDIX 1

1:10 000 maps

The following list shows the 1:10 000 maps included partly or wholly within the area of 1:50 000 Sheet 133 (Aberdaron, including Bardsey Island). Uncoloured dye-line copies of all the maps are available from the British Geological Survey in Keyworth. The list shows the initials of the surveyors and the dates of survey. The surveyors were W Gibbons (solid) and D McCarroll (drift). The mapping of the Monian rocks was initially completed by W Gibbons under BGS supervision in 1976–1979, with final compilation and minor revisions being effected in 1987. The mapping of the Ordovician rocks was undertaken in 1987, with final revisions being completed in 1989.

SH 12 SW	Bardsey and Parwyd	WG, DMcC	1987
SH 12 NW	Mynydd Mawr	WG, DMcC	1987
SH 12 SE	—	WG, DMcC	1987, 1989
SH 12 NE	Aberdaron	WG, DMcC	1987, 1989
SH 13 SE	Porth Widlin	WG, DMcC	1987
SH 22 NW	Rhiw	WG, DMcC	1987, 1989
SH 23 SW	Sarn	WG, DMcC	1987
SH 23 NW	Tudweiliog	WG, DMcC	1987

APPENDIX 2

Geological Survey photographs

Copies of these photographs are available for reference in the Library of the British Geological Survey, Keyworth, Nottingham. Colour or black and white prints and slides can be supplied at the current tariff. The catalogue numbers all fall in the A series.

15001	General view of Bardsey Island from Mynydd Mawr
15002	General view of Bardsey Island from Mynydd Mawr
15003	General view of Aberdaron Bay
15004	Mynydd Ystum. Typical glacial topography of Aberdaron district
15005	Trwyn Maen Melyn
15006	Braich y Noddfa from Braich y Pwll
15008	East of Tyn-Lon. General view across Aberdaron Bay
15009	Trwyn Maen Melyn. Precambrian Gwna Mélange
15010	Trwyn Maen Melyn. Precambrian Gwna Mélange
15011	Trwyn Maen Melyn. Precambrian Gwna Mélange
15012	Trwyn Maen Melyn. Precambrian Gwna Mélange
15013	Trwyn Maen Melyn. Precambrian Gwna Mélange
15014	Trwyn Maen Melyn. Precambrian Gwna Mélange
15015	Braich y Pwll. Precambrian Mélange
15016	Cliffs on SW side of Pen y Cil, looking into Parwyd
15017	Pen y Cil. Dolerite sill and Ordovician rocks
15018	Pen y Cil. Ordovician rocks
15019	SW side Pen y Cil Bau Ogof-eiral. Ordovician rocks
15020	Bau Ogof-eiral. Dolerite sill in Ordovician rocks
15021	Bau Ogof-eiral. Ordovician rocks below dolerite sill
15022	NW side of Parwyd. Ordovician rocks
15023	Bau Ogof-eiral. Ordovician sandstone
15024	Bau Ogof-eiral. Ordovician sediments
15025	Bau Ogof-eiral. Ordovician sediments
15026	Bau Ogof-eiral. Ordovician rocks with thrusts
15027	Porth Ysgo. Ordovician sandstones over dolerite sill
15028	Maen Gwenonwy. Dolerite sill over Ordovician rocks
15029	Maen Gwenonwy. Dolerite sill on Ordovician mudstones
15030	Maen Gwenonwy. Dolerite sill on Ordovician mudstones
15031	Maen Gwenonwy. Structures in dolerite sill
15032	Aberdaron. Quaternary deposits
15033	Aberdaron Bay. Quaternary deposits
15034	Aberdaron Bay. Quaternary deposits
15035	Aberdaron Bay. Quaternary deposits
15036	Aberdaron Bay. Quaternary deposits
15037	Mynydd Penarfynydd. Quaternary glacial topography
15038	Porth Meudwy. Glacial meltwater channel (Quaternary)
15039	Porth Meudwy. Quaternary glacial channel
15040	Porth Meudwy. Quaternary glacial channel
15041	Porth Meudwy. Quaternary glacial channel
15042	Porth Ysgo. Quaternary deposits
15043	Porth Oer. Quaternary deposits
15044	Porth Oer. Quaternary deposits
15045	Porth Oer. Quaternary deposits
15046	Porth Oer. Quaternary deposits
15047	Porth Oer. Quaternary deposits

INDEX

Page numbers in italics indicate figures, bold page numbers indicate tables, P after a page number indicates plates

Aberdaron   *3*, *4*, 39, 42, 54, 55, *58*, 61, 63, 64, *71*, 72, 75, 76
Aberdaron Bay   28, 29, 33, 34, 36, 40, 54, 56, 59, *60*, 67, 69, 71
　coastal drift section   64
Aberdaron Bay Group   *3*, 26–40, 28P, **31**
　palaeontology   29–33
　　pisolitic marker bed   37
Aberdaron Fault   *3*, 55
'Aberdaron Formation'   28, 29
Aberdaron Syncline   53
Acadian deformation   50, 52, 54
AFM plots   20, *21*
Afon Cyll-y-Felin   24, 59, 64, 71
Afon Daron   *3*, 25, 33, 42, 53, 61, 64, 71, 75
　drift sequence   67–68
Afon Daron Gorge   37, 38–39, 55
Afon Fawr   22, 23, 71
Afon Saint Fault   *3*, 54, 55, 76
alleghenyite   77
alluvium   71–72
amphibolite   19
Amplexograptus confertus (Lapworth)   *32*, 33
Anglesey   7, 17, 56, 57
Arenig Series   2, *26*, 28, 29, **30**, 33
Arvon Advances   57
Avalonian rocks   26
Avolonian superterrane   2
Azygograptus eivionicus Elles   29, *32*, 34

Bae y Rhigol   14
banalsite   77
Bardsey Island   *3*, *4*, 6, 7, 8, 9P, 11, 13, 14, 17, 46, 48P, 49, 50, 53, 57, 58, *60*
Baron Hill   41, 42, 43, 70, 76
Barrandia homfrayi Hicks   33
baryte   76
Bau Ogof-eiral   35, 35P, 36, 37P, 38
Benallt   40–41, 54, 55
Benallt Mine   29, 46, 55, 76, 77
Bergamia?   33
Bergamia rushtoni Biozone   **31**
Bod-isaf   71
Bodrydd   71
Bodwrdda   22, 53, *60*, 61, 67, 71
Bohemilla (Fenniops)   *31*, 33
Bohemopyge scutatrix (Salter)   29, 40
Bouma sequences   35
Boundary Thrust   1
Braich Anelog   52

Braich y Pwll   7, 8, *10*, 13, *24*, 49, 49P, 50, 50P, 52
braided stream gravels   68, 69
breccia, lava-limestone   9, 11
Bryn-Chwilog   14
Bryn-mawr   52
building stone   76

Cadlan-uchaf   18
Cadomian rocks   26
Cadwgan   41
Capel Uwch-y-Mynydd   14
Caradocian Epoch   5
Cardigan Bay   57
Careg y Defaid   11
Carn Fadryn   64
Carreg Chwislen   42
Carreg Gybi   42
Carrog   14, 18, 19
Carrog Farm   **20**
'Carw Formation'   28, 29
cataclastites   14
Cefn Enlli   7
clasts, within the Gwna Mélange   7–14
　basalt   9, 11
　granitic   14
　'Gwyddel Beds'   7–9, 13
　limestone   11, 13
　quartzite   11
　red mudstone   13–14
Clip Lava   42, 44
Clip y Gylfinhir   41, 44
Clynnog Fawr moraine   58, 64, 75, 76
Cnemidopyge salteri (Hicks)   29
Coedana Granite   17
copper   76
Craig Fael   58
Crugan Bâch   17, 18, 19, 51
Cryptograptus tricornis schaeferi Lapworth   33
Cyclopyge   29, 34
Cyclopyge aff. *grandis* (Salter)   31P, 31
Cyclopyge grandis brevirhachis Fortey and Owens   31, 40
Cyclopyge grandis grandis (Salter)   29, 31, 31P, 39
cymrite   77

'Daron Cherts'   39
Daron Fault   *3*, 23, 25, 27, 28, 29, 36, 42, 53, 54, 55
Devensian   5
　see also Late Devensian
diamict
　lower   65, 70
　upper   65–67, 68, 70
Didymograptus artus Biozone   **31**, 33
Didymograptus cf. *extensus linearis* Monsen   31, 40
Didymograptus deflexus Biozone   **31**
Didymograptus (D.) artus Elles and Wood   33
Didymograptus (D.) spinulosus Perner   *32*, 33
Didymograptus distinctus Harris and Thomas   31, 40
Didymograptus hirundo Salter   31, *32*, 40

Didymograptus hirundo Biozone   **31**
Didymograptus aff. *nicholsoni* Lapworth   33
Didymograptus nitidus Biozone   **31**
Didymograptus praenuntius Tornquist   34
Didymograptus uniformis lepidus Ni   31, 40
Dimlington advance   73
Dimlington Stadial   57
Dinas Bâch   7, 9, 57
Dinas Fawr   7, 9, 11P, 13, 51, 57
Dindymene cf. *didymograpti* (Whittard)   33
Dionide levigena Fortey and Owens   33
Dionide levigena Biozone   **31**
Dionide turnbulli Whittington   33
diorite, Sarn Complex   18–19
　XRF analyses   **20**
Diplograptus ellesi Bulman   33
dolerite intrusions   *3*, 5, *10*, 42–43, 48P
Drepanocladus revolvens   74
drift-filled valleys   59, 61
drift, Irish Sea   61, 75
drift sequences, environments of deposition   74
Dwyrhos   29
Dwyrhos Quarry   29
dykes   41
　olivine dolerite (Tertiary)   5, 41, 46–47

Ebolion   33, 34, 36
Eoglyptograptus   33
erosion, periglacial   72
etchplanation   72

faults, post-Ordovician   54–55
Fennian Stage   2, 5, *27*, 28, 31–33, **31**, 37, 39, 40
flow tills   66, 68, 69
folds, post-Ordovician   53–54
Footwall Sill   42, 55
Furcalithus   29, 34
Furcalithus aff. *sedgwickii* (Salter)   29
Furcalithus radix Biozone   **31**

gabbro, Sarn Complex   19
　mineralogy   19
　XRF analysis   **20**
Gallt y Mor Sill   33, 42, 43, 43P
Garn Boduan   64
Garreg Fawr   76
gibberulus Biozone   **31**
Gibula umbilicalis   71
glacial deposits, late Devensian   61, 63–69
glacial grooves   57–58
glacial sediments, interpretation of   69–70
glacial stages   *58*
glacial striations   57–58, 59P, 69
glaciation, Quaternary   57–61, 73–75
Glanllynau   73
Graig Fael   18, 19, **20**, 76
granite, Sarn Complex   17–18
granodiorite, XRF analysis   **20**
granophyllite   77

gravels 76
  braided stream 68
'The Great Quartzite' 11, 24
Gwna Mélange 1, 2, *3*, 6P, 6–14, *10*, *15*, *24*, 25, 26, 49, 50, 51, 52, 53, 54, 59P
  clasts within 6P, 7–14, 9P, 11P, 13P, *15*, 53P
'Gwyddel Beds' 2, 7–9, 8P, *10*, 13, 14, 14P, 49P, 50P, 51, 52
'Gwynedd Readvance 57, 58
Gymnostomix gibbsi (Salter) 29
Gymnostomix gibbsii Biozone **31**

Hanchungolithus cf. *primitivus* (Born) 29P, 31
head 62–63, 70
Hellwyn 7
Hen Borth 58
Hendre-Uchaf 54
Henllwyn 57
hirundo Biozone 31

Iapetus Ocean 5
ice-flow directions 56, 58, *60*
igneous rocks, intrusive 42–47
inselbergs 56, 72
interglacial raised beach 61
intrusions
  dolerite 5, 42–43
  layered 5, 41, 44–46
  see also dykes; sills
Irish Sea Basin 72, 73, 75
Irish Sea drift 61, 75
Irish Sea glacier 69, 75
Irish Sea ice 57, 64
Irish Sea till 62
ironstone, pisolitic 37
Isograptus caduceus gibberulus (Nicholson) 31, 32, 40
Isograptus gibberulus Biozone **31**

jacobsite 77
Janograptus 33
jasper 7, 76
'junction bed' 33, 34

kettle holes 64, 70

landform assemblages 64
landscape evolution, periglacial 56
landslips 71
Late Devensian Glacial deposits 61, 63–69
Late Devensian ice 64, 73
  extent of 73
lime, agricultural 76
Lingulella? 33
Llanfaelrhys 40, 44
Llangwnnadl 9, 14, 18, 19, 22
  stream section *22*, 23
Llanvirn Series 2, 5, *27*, 28, 29, 31, 33, 37, 39, 41, 44
Llŷn Shear Zone 2, *3*, *4*, 13, 21–25, *22*, *24*, 25P, 35P, 49, 50, 54, 72
lodgement till 69
Lonchograptus *32*, 33

lower diamict 64, 70
Lower Palaeozoic, Welsh Basin 2, 5

Maen Aber-dywyll 7
Maen Gwenonwy *3*, 28P, 29, 34, 62
Maen Gwenonwy Sill 42
Maen Melyn Lleyn 52
Main Anglesey Advance 57, 58
manganese 76
marine planation 72
Meillionydd 17, 19, 51
Meillionydd Farm 18
Mellionedd Farm **20**
melt-out till 65
Menai Strait Fault System *4*, 72
Merlinia rhyakos Fortey and Owens 29, 34
Merlinia rhyakos Biozone **31**
Merlinia selwynii (Salter) 29, 30P, 31, 34, 40
Merlinia selwynii Biozone **31**
Methlem Farm 61
Microparia broeggeri (Holub) 31, 40
Microparia sp. 30P
Mochras Borehole 72
Monian rocks 1, 2, 23, 25, 35, 49, 51
morainic deposits 64
  see also Clynnog Fawr moraine
Moridunian Stage 2, *27*, 28, 29, **31**, 33, 34, 40
Mountain Cottage Quarry 17, 23, 28
mylonite 14, 23, 25
Mynydd Anelog 7, 8, 9, 13, 46, 51–52, 55, 58, 76
Mynydd Bychestyn *24*, 25P, 51
Mynydd Carreg 7, 24, 55, 57P, 59, *60*, 71, 76
Mynydd Cefnamwlch 14, 17, 64
Mynydd Enlli 7, 8, 28P, 53, 58
Mynydd Mawr 7, 8, 8P, 10, 11, 14, *24*, 46, 52, 53, 58, 76
Mynydd Mawr Syncline 51
Mynydd Penarfynydd 2, 5, 28P, 41, 44, 46, 59, *60*, 71, 76
  layered intrusion 42, 44–46, 46P, 47P
Mynydd Rhiw 41, 43, 44, *60*, 64, 70, 77
Mynydd y Gwyddel 7, 8, *10*, 14, *24*, 52
Mynydd Ystum *3*, 22, 24, 51, 54, 55, 59, *60*, 61, 64, 71

Nant Mine 40, 55, 76, 77
Nant y Gadwen 29, 40, 44, 54, 55, 57
Nant y Gadwen Valley 61
Nant-y-Carw 28
Nefyn 57
North Atlantic Ocean 5, 47
Northern ice 57
Novakella? 29, 34

Ogof Ddeuddrws 29, 34
Ogof Hir 46
Ogof Lleuddad 29, 33, 34, 42, 55
Ogof Newry 8
Ogof Pren-côch 8, 55
olistostromes 1
Ordovician rocks 2, 5, 27–41, 49

graptolites *32*
Oxygen Isotope Stages 57, 61, 73

Pandy-Bodwrdda 38, 42, 43
Pared Llech-ymenyn 11, 13, 13P, 23, 24, 52
Pared Thrust 24, *24*
Parwyd 1, *3*, 17, 21, *24*, 28, 33, 34, 54, 55
Parwyd Bay 29
Parwyd Fault *3*, 55
Parwyd fault system 54–55
Parwyd Gneisses 3, 21, 23, *24*, 35P
Parwyd Thrust 23, *24*, 53
Patella vulgata 71
Paterula 33
Pen Cristin 53, 57
Pen y Cil *3*, *24*, 35, 42, 58
Pen y Cil Sill 42, 54
Pen y Gopa 17, 18, **20**
Pen-Cae 25
Pen-yr-Orsedd 7, 9, 51
Penarfynydd 33
Penarfynydd Farm 29, 41
Pencaerau 64
peneplains 72
'Penmynydd Zone of Metamorphism' 22, 23
pennantite 77
Penrhyn Colmon 69
Penrhyn Gogor 57
Penrhyn Mawr 1, 22, 25, 33, 61, 69
Penrhyn Mawr Farm 55
Penrhyn Melyn 7
Penrhyn Nefyn 23
peperites 5, 42, 43
periglacial erosion 73
periglacial landscape evolution 56
phosphatic rocks 33, 34, 38P
Phycodes 33
pillow lavas 5, 6, 9, *11*, 24, 28P, 42, 43, 56
Platycalymene 39
Platycalymene tasgarensis tasgarensis Shirley 30P, 31
Pont Afon Saint 64
Pont Cyll-y-felin 9, 24
Pont Llangwnnadl 23
Pont yr Afon Fawr 71
Porth Bâch 7
Porth Cadlan 28, 33, 34, 53
Porth Cloch *3*, 54, 55
Porth Cloch Anticline 53, 54
Porth Colmon 11, 46, 50, 51, 57, *60*, 69
Porth Dinlleyn 57
Porth Felen 6, 7, *10*, 13, 13P, 14P, *23*, 51, 71
Porth Gwylan 50
Porth Hadog 53
Porth Iago 51
Porth Llanllawen *60*, 61, 64
Porth Llawenan 69
Porth Llong 8, 50
Porth Llydan 51, 61
Porth Llyfesig 51
Porth Meudwy *3*, 33, 34, 36, 38, 38P, 39P, 42, 53, 54, 55, 56, 58, 59

Porth Meudwy Formation 24P, 27, *27*, 29, **31**, 34–36, 35P, 37P
  Facies A 35
  Facies B 36
  Facies C 36
  Facies D 36
Porth Meudwy Member 29
Porth Oer 6, 9, 42, 50, 51, 56, 57, *60*, 61, 62P, 64, 68, 69, 71
  raised beach 61
Porth Simdde 33, 53, 56, 59, 62, 76
Porth Solfach 14, 57, 58, *60*
Porth Tŷ-mawr 11, 50
Porth Wen Bâch 50
Porth Widlin 7, 9, 49, 51
Porth y Pistyll 56, 61
Porth y Wrâch 8, 51
Porth Ysgaden 7, 11, 51, 61, 69
Porth Ysgo 29, 33, 42, 43P, 44, 53, 55, 63, 63P, 69
Porth Ysgo Fault 43P
Porthorion 7, 9, 13, 51, 71
postglacial raised beach 71
preglacial landscape evolution 56, 72–73
Pricyclopyge binodosa eurycephala Fortey and Owens 31, 40
Pseudisograptus angel Jenkins 31, 40
Pseudisograptus cf. *dumosus* (Harris) 31, 40
Pseudoclimacograptus 33
Pseudotrigonograptus minor (Mu and Lee) 40
Psilacella *30*, 30
Pwll Diwaelod 64
Pwll Hwyaid 7
pyrochlorite 77

Quaternary 5, 56–57
  deposits 61–71
  glaciation 57–61
  history 72–75

raised beach platform 56, 57
raised beaches
  Porth Oer (interglacial) 61
  postglacial 71
  raised rock platform 72
rare earth elements (RRE) 19
Rb-Sr isotopic ages 21

Rhiw Boreholes 1A and 1B 40
Rhiw Igneous Complex 44
rhodonite 77
Rhoshirwaun *60*
Rhos Hirwain Syenite 17
Rhydlios 64

St Tudwal's Peninsula 57
sand 76
Sarn Complex 2, *3*, 14–21, 25, 49, 51
  geochemistry 19–21
'Sarn Formation' 28, 29, 34
Sarn Granite 14, 17, 18
  XRF analyses **20**
scree 62–63, 63P, 68, 69
Segmentagnostus hirundo (Salter) 29
Shumardia *30*, 31
Shumardia gadwensis Fortey and Owens 29
sills 35, 42
'slaty marble mélange' 11
spessartine 77
Stapeleyella abyfrons Biozone **31**
Stepeleyella murchisoni (Slater) *30*, 31
structures
  post-Ordovician 53–55
  pre-Ordovician 49–53
syenite 17

tectonic setting, Precambrian rocks 25–26
Teichichnus 34
tephroite 77
terranes, suspect 25
Tetragraptus reclinatus Elles and Wood 31
till
  flow 66, 68, 69
  Irish Sea 62
  lodgement 69
  melt-out 65
tonalite, Sarn Complex 17–18
  XRF analyses **21**
topography, preglacial 56
Traeth Penllech 50, 51, 64
Tremadoc Bay 56
Trwyn Bychestyn *3*, 11, 21, 23, 42, 46, 50, 51, 62
Trwyn Cam *3*, 33, 34, 51, 54, 55
Trwyn Cam Member *27*, 34, 36

Trwyn Cam Syncline 53, 54
Trwyn Dihirid 7
Trwyn Gareg-lwyd 50, 51
Trwyn Glas 7, 51
Trwyn Maen Melyn 6, 6P, 7, *10*, 11, 42, 50, 52
Trwyn Talfarach 44, 46P, 64, 76
Trwyn y Gwyddel 52
Trwyn y Penrhyn *3*, 34, 39, 44, 55, 64, 76
25-foot platform *see* raised beach platform
Ty Canol Dyke 46
Ty Fwg 64
Ty Nant 71
Ty-isaf 71
Ty-tan-yr-allt 54
Tyddyn Meirion 40, 77
Ty'n-lôn-fach 64

Uncinisphaera? 29
upper diamict 65–67, 68, 70
Uwchmynydd 7, 51, 59, *60*, 64, 71

Welsh Basin 2, 3, *4*
Welsh ice 57
Whitlandian Stage 2, *27*, 28, 29, **31**, 34, 40
Wig 1, *3*, 23, 25, 29, 33, 53, 54, 55, 62, 64, 68
Wig Bâch 28, 29, 33, 34
Wig Bâch Formation *24*, 27, *27*, 28P, 29, **31**, 33–34
  Trwyn Cam Member 34, 36
  Wig Member 34
Wig Fault 23, 33, 34, 55
Wig Member *27*, 34

XRF analyses 20
  Penarfynydd layered intrusion **45**
  Sarn Complex **20**

Ynys Enlli *see* Bardsey Island
Ynys Gwylan-bâch 28P
Ynys Gwylan-fawr *3*, 28P, 39, 42, 43
Ynys Piod 34, 76
Ystol-helig-bâch 14
Ystum Fault *3*, 55

BRITISH GEOLOGICAL SURVEY

Keyworth, Nottingham NG12 5GG
(0602) 363100

Murchison House, West Mains Road, Edinburgh
EH9 3LA   031-667 1000

London Information Office, Natural History Museum
Earth Galleries, Exhibition Road, London SW7 2DE
071-589 4090

The full range of Survey publications is available through the Sales Desks at Keyworth and at Murchison House, Edinburgh, and in the BGS London Information Office in the Natural History Museum Earth Galleries. The adjacent bookshop stocks the more popular books for sale over the counter. Most BGS books and reports are listed in HMSO's Sectional List 45, and can be bought from HMSO and through HMSO agents and retailers. Maps are listed in the BGS Map Catalogue, and can be bought BGS approved stockists and agents as well as direct from BGS.

The British Geological Survey carries out the geological survey of Great Britain and Northern Ireland (the latter as an agency service for the government of Northern Ireland), and of the surrounding continental shelf, as well as its basic research projects. It also undertakes programmes of British technical aid in geology in developing countries as arranged by the Overseas Development Administration.

The British Geological Survey is a component body of the Natural Environment Research Council.

HMSO publications are available from:

HMSO Publications Centre
(Mail, fax and telephone orders only)
PO Box 276, London SW8 5DT
Telephone orders 071-873 9090
General enquiries 071-873 0011
Queueing system in operation for both numbers
Fax orders 071-873 8200

HMSO Bookshops
49 High Holborn, London WC1V 6HB
(counter service only)
071-873 0011   Fax 071-873 8200
258 Broad Street, Birmingham B1 2HE
021-643 3740   Fax 021-643 6510
33 Wine Street, Bristol BS1 2BQ
0272-264306   Fax 0272-294515
9 Princess Street, Manchester M60 8AS
061-834 7201   Fax 061-833 0634
16 Arthur Street, Belfast BT1 4GD
0232-238451   Fax 0232-235401
71 Lothian Road, Edinburgh EH3 9AZ
031-228 4181   Fax 031-229 2734

HMSO's Accredited Agents
(see Yellow Pages)

And through good booksellers